Exploring Science

Content Consultants
Randy L. Bell, Ph.D.
Malcolm B. Butler, Ph.D.
Kathy Cabe Trundle, Ph.D.
Judith S. Lederman, Ph.D.

On the Cover
The Colorado River curves at Horseshoe Bend near Page, Arizona.

Physical Science

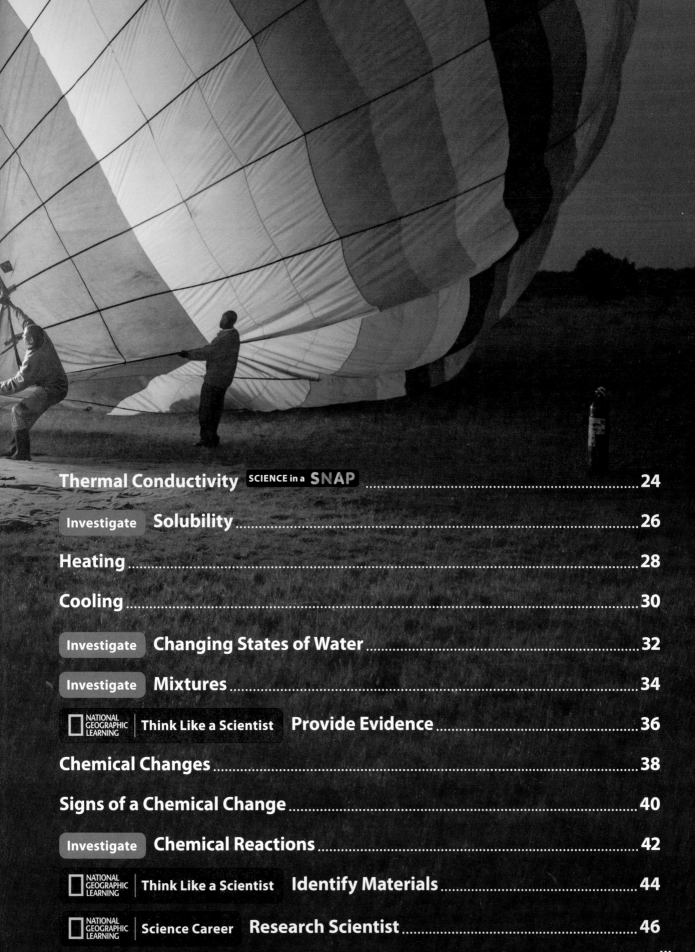

Life Science

Earth Science

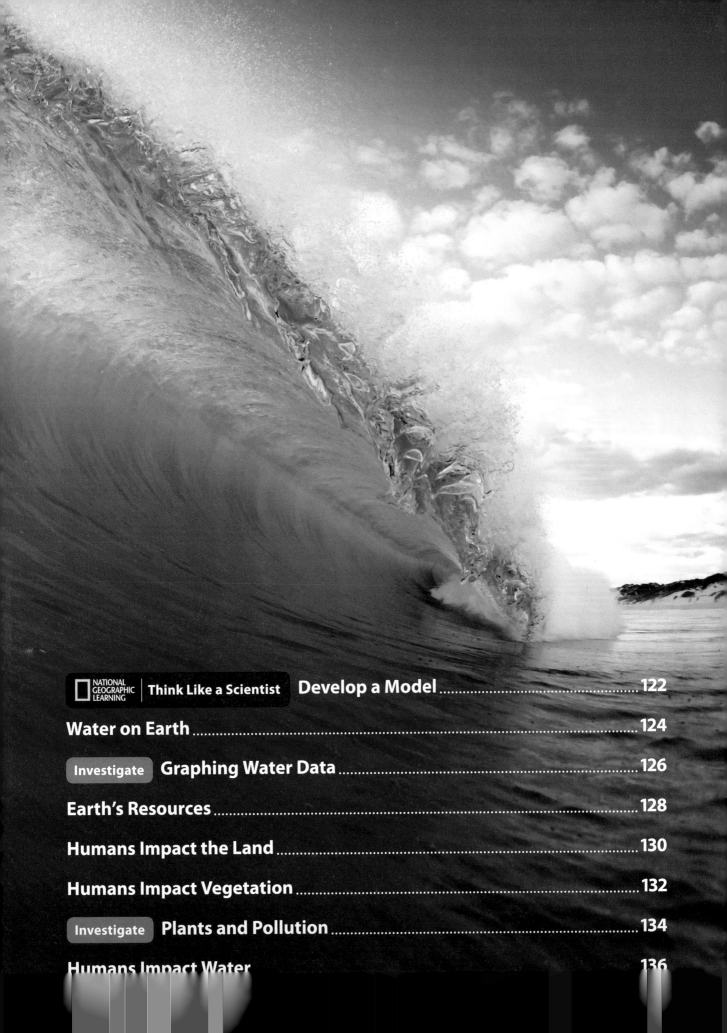

Earth Science (continued)

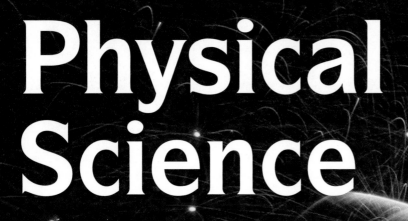

Physical Science

Structure and Properties of Matter

When chemicals in fireworks burn,
they produce light of different colors.

3

Matter

Can you identify the matter in this picture? Here's a hint—everything you see in the picture is made of matter. The sandcastle is made of matter and so is the water. Even the air is made of matter. **Matter** is anything that has mass and takes up space.

Mass is the amount of "stuff" in an object. Imagine holding two balls of the same size. One is made of foam, and the other is made of the metal lead. Which would feel heavier? The lead ball, of course!

This sandcastle is matter. It takes up space on the beach. If you could put it on a scale, you would see that it also has mass.

NEXT GENERATION SCIENCE STANDARDS | DISCIPLINARY CORE IDEAS
PS1.A: Structure and Properties of Matter
Matter of any type can be subdivided into particles that are too small to see, but even then the matter still exists and can be detected by other means. (5-PS1-1)

That's because the metal ball has more matter in it, and so, more mass.

Matter of any type can be broken down into smaller parts. A sandcastle can be broken down into smaller grains of sand. And a grain of sand can be broken down into even smaller particles, which are too small to see. But even these tiny particles are matter.

The sandcastle is made up of thousands of grains of sand.

Even though this grain of sand has been magnified 100 times, you still can't see the particles that make it up. They are too small to see.

My science notebook

Wrap It Up!

1. **Define** What is matter?

2. **Infer** Do sand particles have mass? Explain.

States of Matter

Matter can be classified by its state. Solids, liquids, and gases are all physical **states of matter.** Each state has specific characteristics. For example, **solids** have a definite shape. A brick is a solid.

Liquids, such as milk in a bottle, take the shape of their containers. Liquids do not necessarily fill a container completely.

If you have ever seen a balloon floating in the air, you have seen an object filled with gas. **Gases** have no definite shape. Gases spread out to completely fill a closed container.

Solid This sailboard is a solid. The particles that make up a solid are close together and vibrate, or "jiggle," in place.

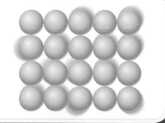

Liquid The ocean is a liquid. Particles of a liquid are farther apart. The particles move more freely than the particles of a solid.

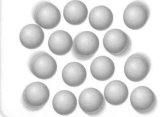

NEXT GENERATION SCIENCE STANDARDS | DISCIPLINARY CORE IDEAS
PS1.A: Structure and Properties of Matter
Matter of any type can be subdivided into particles that are too small to see, but even then the matter still exists and can be detected by other means. A model showing that gases are made from matter particles that are too small to see and are moving freely around in space can explain many observations, including the inflation and shape of a balloon and the effects of air on larger particles or objects. (5-PS1-1)

Gas The air that surrounds our planet is a gas. Moving air, or wind, can push objects such as this sail. The particles that make up a gas move around freely.

Feel the Air

1 Observe a balloon. Press on it. Stretch it. Record your observations.

2 Now blow some air into the balloon. Tie it off so the air cannot escape. Observe the shape and feel of the balloon. Record your observations.

? How do your observations provide evidence that air is made of matter?

My science notebook

Wrap It Up!

1. **List** What are three physical states of matter?

2. **Classify** Find examples of solids, liquids, and gases in your classroom. Explain how you classified each of the objects as a solid, a liquid, or a gas.

3. **Apply** Is honey a solid, liquid, or gas? How do you know?

Investigate

Matter

? **How can you detect materials that have dissolved in water?**

As salt is stirred into water, it seems to disappear. But the salt is not gone—it dissolves. When a solid **dissolves** in a liquid, the tiny particles that make it up become evenly mixed into the liquid. How do you know the salt is still there? If the liquid **evaporates,** or changes state from a liquid into a gas, the solid is left behind. In this investigation, you will use evaporation to separate salt from salt water.

Materials

salt	water in a cup	spoon
dropper	black construction paper	hand lens

NEXT GENERATION SCIENCE STANDARDS | DISCIPLINARY CORE IDEAS
PS1.A: Structure and Properties of Matter
Matter of any type can be subdivided into particles that are too small to see, but even
then the matter still exists and can be detected by other means. (5-PS1-1)

8

1 Pour a spoonful of salt into the water. Stir until you can no longer see the salt. Record your observations.

2 Use the dropper to place three separate drops of salt water onto black paper.

3 Let the water evaporate for one hour or more.

4 Use the hand lens to examine the three areas of the paper where you dropped the salt water. Record your observations.

The smaller the crystals of salt are, the more quickly they dissolve in water.

Wrap It Up!

1. **Describe** What did you observe on the paper after the water evaporated?

2. **Explain** How do your results provide evidence that matter is made of particles too small to see?

Develop a Model

You have observed evidence that particles of matter exist even when they are not visible. When you blew up a balloon, particles of air pushed on the inside of the balloon and filled it up. When you let salt water evaporate, particles of salt were revealed. Now it's your turn. Imagine that you want to explain what you have learned to a friend. What model can you use to explain that matter is made of small particles too small to be seen?

1. **Construct an explanatory model.** *my science notebook*
   Study the materials your teacher makes available to you. Think about how you could use these materials to show evidence that matter is made of particles too small to be seen. Draw and label a picture of your model. Explain how your model will work.

2. **Conduct an investigation.**
   After your teacher approves your design, build your model. Conduct an investigation that tests your model. Collect data from your test.

3. **Analyze results and revise your model.**
   Study your data. Does your model do what you want it to do? Revise your model and retest until you are satisfied with how it works.

4. **Share your model.**
   Use your model to explain to a partner that matter is made of particles too small to be seen. Does your partner "get" it? If not, work together to further improve your model. Then share your final model with the class.

Hot gas particles fill these balloons and allow them to float.

Properties of Matter

Your friend asks you to close your eyes. In your hand, he places a large, polished rock. Even before you open your eyes, you know the object is a rock based on your sense of touch. The rock's large mass and hard, smooth surface give it away. Scientists can measure the properties of the rock more precisely to identify exactly what type of rock it is.

Physical **properties** are observable characteristics of a material that identify the material. Look at the different objects on these pages. Each has observable characteristics. Often those characteristics make the objects practical for a specific use.

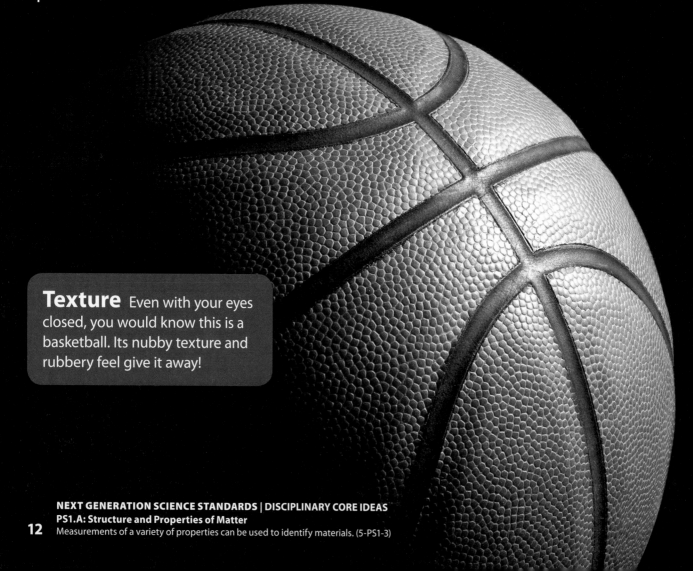

Texture Even with your eyes closed, you would know this is a basketball. Its nubby texture and rubbery feel give it away!

NEXT GENERATION SCIENCE STANDARDS | DISCIPLINARY CORE IDEAS
PS1.A: Structure and Properties of Matter
12 Measurements of a variety of properties can be used to identify materials. (5-PS1-3)

Color and Shape
When you search for your purple helmet, you are using color to describe an object. Its round shape allows it to fit snugly on your head.

Hardness
Could you use a stuffed toy to drive a nail into wood? Of course not! You need something hard and strong, such as this hammer.

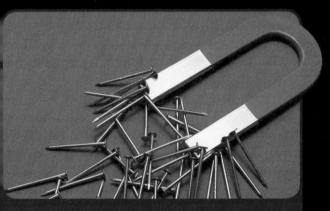

Magnetism
The iron in these nails is attracted to the magnet. Magnets also attract cobalt and nickel.

Reflectivity
What do mirrors and these shiny pots have in common? They all reflect light in a way that allows you to see an image.

Solubility
The property of solubility allows you to mix up a cold glass of grape drink. The powder dissolves in the water.

My science notebook

Wrap It Up!

1. **List** Name six physical properties that can be used to identify matter.

2. **Apply** Choose an object in your surroundings. Describe its physical properties.

Hardness is a measure of how resistant a material is to scratching, bending, or denting. Hardness is measured on a scale that ranks materials from very soft to very hard. For example, scientists use hardness as a way to identify minerals. The chalk this artist is using to create the image on the sidewalk contains a soft mineral. It rubs off on the

The pavers are harder than the chalk. When scratched across the pavers, the softer chalk leaves a mark.

NEXT GENERATION SCIENCE STANDARDS | DISCIPLINARY CORE IDEAS
PS1.A: Structure and Properties of Matter
14 Measurements of a variety of properties can be used to identify materials. (5-PS1-3)

Magnetism

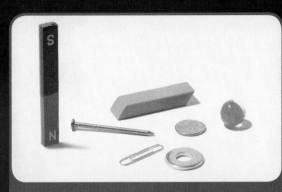

1 Observe a group of objects. Predict which objects will be attracted to a magnet. Sort the objects into two piles—ones you think are magnetic and ones you think are not magnetic.

2 Use the magnet to try to pick up each object. Record your observations.

? **Did your results support your predictions? Did any of your results surprise you?**

Maglev trains do not have wheels that roll on rails. Magnetism is used both to levitate, or lift, the train above the track and to move the train forward.

Wrap It Up!

1. **Identify** Give an example of an item that is magnetic. What type of metal does it most likely contain?

2. **Explain** How can the property of magnetism be tested?

Electrical Conductivity

Electrical conductivity is another property of matter. **Electrical conductivity** is a measure of how well electricity can move through a material. Good conductors of electrical energy, or **electrical conductors,** allow electricity to flow easily. Metals such as copper, gold, silver, and iron are good electrical conductors. Copper is commonly used to make electrical wires because it is such a good conductor.

Electrical insulators cover the lights and pumps in this pool. The lights and pumps are powered by electricity.

NEXT GENERATION SCIENCE STANDARDS | DISCIPLINARY CORE IDEAS
PS1.A: Structure and Properties of Matter
Measurements of a variety of properties can be used to identify materials. (5-PS1-3)

Because electricity can be dangerous, it is important to protect people from it. An **electrical insulator** is a material that slows or stops the flow of electricity. Plastic, rubber, wood, and glass are good electrical insulators.

ELECTRICAL CONDUCTORS

Copper wires carry electricity in power lines and electrical plugs.

Gold carries electricity in some parts of a computer.

ELECTRICAL INSULATORS

Plastic coating on wires helps to prevent electrical shocks.

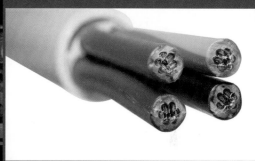

Glass electrical insulators prevent electricity from reaching a person working on power poles.

My science notebook

Wrap It Up!

1. **Contrast** What is the difference between an electrical conductor and an electrical insulator?

2. **Recall** From what you've learned, what are two different properties of iron?

3. **Apply** Explain why electrical gloves are made of rubber.

21

Investigate
Electrical Conductivity

? **Which materials conduct electricity?**

You will never see electrical wires made of wood. Wood just isn't a good conductor. Wood is, however, a good insulator. How well a material conducts electricity is a physical property of the material. In this investigation, you'll test the ability of different materials to conduct electricity.

Materials

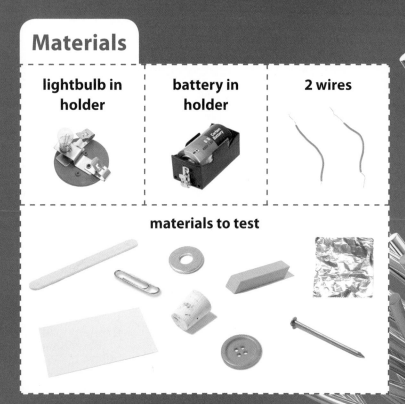

lightbulb in holder	battery in holder	2 wires

materials to test

The plastic around the copper wire is a good insulator. The copper wire is a good conductor.

NEXT GENERATION SCIENCE STANDARDS | DISCIPLINARY CORE IDEAS
PS1.A: Structure and Properties of Matter
22 Measurements of a variety of properties can be used to identify materials. (5-PS1-3)

1 Examine the materials to test. Predict which materials will conduct electricity. Record your predictions in your science notebook.

2 Attach the first wire to one end of the battery holder by connecting it to the metal piece. Attach the other end of the wire to the bulb holder. Attach the second wire to the other end of the battery holder.

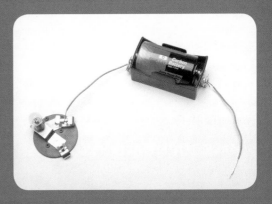

3 Wrap the end of the free wire around the nail. Touch the end of the nail to the open end of the bulb holder. Record your observations.

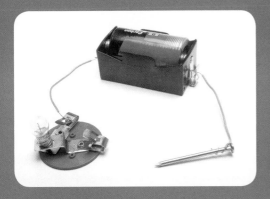

4 Use the free end of the wire and bulb holder to test the rest of the materials. Record your observations.

Wrap It Up!

1. **Explain** Did your results support your predictions? Explain.

2. **Classify** Identify each material as an electrical conductor or an electrical insulator.

property that can be used to identify materials.

Good conductors of thermal energy, or **thermal conductors,** allow thermal energy to flow easily through them as heat. Metals such as copper, aluminum, and iron are good thermal conductors.

The metal rod on the thermometer conducts thermal energy.

Iron is a good thermal conductor. Many pots and pans are made of metals, including iron.

This spatula is made of wood. Wood is a good thermal insulator.

A stove burner provides a heat source to cook food.

NEXT GENERATION SCIENCE STANDARDS | DISCIPLINARY CORE IDEAS
PS1.A: Structure and Properties of Matter
24   Measurements of a variety of properties can be used to identify materials. (5-PS1-3)

Cloth, wood, and rubber objects do not conduct thermal energy well. These materials are **thermal insulators.** We can use these materials to protect us from hot objects, such as a pot on a stove. Glass, plastic, and leather are also good thermal insulators.

Pot holders are made of cloth. Cloth is a good thermal insulator.

Test Thermal Conductors

1 Predict which spoon will be a better thermal conductor. Record your predictions.

very warm

2 Test your prediction. Place the spoons in a cup of very warm water. Feel the stems of the spoons after a few minutes. Record your observations.

? **How did the thermal conductivity of the spoons differ?**

My science notebook

Wrap It Up!

1. **Contrast** What is the difference between a thermal conductor and a thermal insulator?

2. **Classify** Identify the following materials as thermal conductors or thermal insulators: a wooden spoon, an iron frying pan, a plastic spatula, a steel fork.

3. **Apply** Explain why ceramic cups for drinks like coffee, tea, or cocoa usually have handles.

Solubility

? **Which materials dissolve in water?**

When sugar is stirred into lemonade, the sugar dissolves. The mixture of sugar and lemon juice is called a **solution.** In a solution, dissolved particles are distributed evenly and you can no longer see them. The ability of one substance to dissolve in another, or **solubility,** is a physical property of matter. In this investigation, you'll test the solubility of materials in water.

Materials

| 4 cups of water | sand |
| plastic spoon | salt | lemon juice | vegetable oil |

NEXT GENERATION SCIENCE STANDARDS | DISCIPLINARY CORE IDEAS
PS1.A: Structure and Properties of Matter
26 Measurements of a variety of properties can be used to identify materials. (5-PS1-3)

1 Predict what will happen when sand is added to water. Add a half spoonful of sand to a cup with water. Stir the water for about 30 seconds. Record your observations in your science notebook.

2 Predict what will happen when salt is added to water. Add a half spoonful of salt to a cup with water. Stir the water for about 30 seconds. Record your observations in your science notebook.

3 Predict whether lemon juice is soluble in water. Then predict whether vegetable oil is soluble in water. Record your predictions.

4 Pour 25 mL of lemon juice into the third cup of water. Stir for about 30 seconds. Repeat using vegetable oil and the fourth cup of water. Record your observations.

What do you think lemonade tastes like if you take a sip before the sugar dissolves?

 My science notebook

Wrap It Up!

1. **Predict** Did your results support your predictions? Explain.

2. **Classify** Identify each material used in this investigation as soluble or insoluble in water.

Heating

Matter can change states when it is heated. For example, when ice is heated to its **melting point,** it melts into water. When water is heated to its **boiling point,** it becomes a gas called water vapor. Whether water is a solid, liquid, or gas, it is still water. These changes in state from a solid to a liquid and back again are known as **physical changes.** Physical changes, such as boiling water or cutting paper with a pair of scissors, do not change the material into a different one.

At its boiling point, water begins to change from a liquid to a gas. This change occurs at a temperature of 100°C (212°F).

NEXT GENERATION SCIENCE STANDARDS | DISCIPLINARY CORE IDEAS
PS1.A: Structure and Properties of Matter
The amount (weight) of matter is conserved when it changes form, even in transitions in which it seems to vanish. (5-PS1-2)

As water changes and rises from a boiling pot, it seems to vanish. But the water is still there, just in a different form. Even when matter changes state, the amount of matter stays the same. This principle is called the **conservation of matter.**

The melting point of water is 0°C (32°F). If the temperature of frozen water rises to 0°C (32°F), it begins to change into a liquid.

Wrap It Up!

science

1. **Identify** Boiling and melting points a properties of matter. What are the boili melting points of water?

2. **Cause and Effect** How does boiling water's state of matter?

Cooling

Just as heating causes matter to change states, so does cooling. **Condensation** is the change from a gas to a liquid. As the water vapor over a pot of boiling water cools in the air, the water begins to condense. That produces the fog-like steam you see over the pot. Water vapor itself is invisible. But steam is actually tiny drops of liquid water! When the water vapor changes to steam, the amount of matter is conserved.

It hasn't rained, yet this spiderweb is lined with beads of water. The cooler air around the web caused the water vapor in the air to **condense** on the web.

NEXT GENERATION SCIENCE STANDARDS | DISCIPLINARY CORE IDEAS
PS1.A: Structure and Properties of Matter
The amount (weight) of matter is conserved when it changes form, even in transitions in
which it seems to vanish. (5-PS1-2)

Water condenses when moist, warm air comes in contact with a cooler surface. That's why the bathroom mirror fogs up when you take a shower.

When liquid water is cooled to 0°C (32°F), it begins to freeze. Sometimes the air cools so quickly that water vapor from the air condenses and quickly freezes. Little bits of ice called frost form. Frost is common in areas that have a lot of moisture in the air and often occurs at night, when temperatures drop. When water vapor condenses and freezes, the amount of matter is conserved. Frost can form on the insides of windows in the winter, when the water vapor in the warm inside air comes in contact with a cold window.

The water on the outside of this cold glass didn't escape from the inside of the glass. It condensed from the water vapor in the surrounding air.

The frost on this glass is really water from the air that condensed and then froze.

Wrap It Up!

1. **Define** What is condensation?

2. **Explain** How can a window that is not wet become covered with frost?

3. **Summarize** Make a diagram to show water's three states of matter and its change from one to another. Include the labels: *ice, water, water vapor, condensation, melting, freezing.*

Changing States of Water

? **How is matter conserved when water changes state?**

You're already familiar with water in all three of its states: solid, liquid, and gas. Even when water changes from one form to another, its mass is conserved, or stays the same. However, some physical properties of water *do* change when it changes state. In this investigation, you'll observe some of these changes.

Materials

2 resealable bags	tape	balance
water	graduated cylinder	gram masses

NEXT GENERATION SCIENCE STANDARDS | DISCIPLINARY CORE IDEAS
PS1.A: Structure and Properties of Matter
The amount (weight) of matter is conserved when it changes form, even in transitions in which it seems to vanish. (5-PS1-2)

32

1 Label 2 plastic bags *Bag 1* and *Bag 2*. Use a graduated cylinder to measure 100 mL of water. Pour the water into the bag. Seal the bag. Repeat with the other bag.

2 Use the balance and gram masses to measure the mass of each bag. Record your observations in your science notebook.

3 Predict what will happen when you freeze the water. Record your predictions. Place the bags in a freezer. The next day, take the bags out of the freezer. Measure the mass of the bags. Turn the bags in different directions and measure the mass again. Record your observations.

4 Place the bags in sunlight. Open Bag 1 only. Predict what will happen to the water in the bags after 3 days. Observe the bags every day for 3 days. Measure the mass of the bags after 3 days. Record your observations.

Wrap It Up!

1. **Predict** Did your results support your predictions? Explain.

2. **Compare and Contrast** Which properties of water stayed the same after cooling? Which properties changed?

3. **Infer** Explain the differences in the bags after step 4.

4. **Draw Conclusions** How do your findings demonstrate the conservation of matter?

Mixtures

? **How is matter conserved when baking soda and water mix?**

You have found that the amount of matter stays the same, even when it changes from one state to another. What happens when one material is mixed with another? Does the amount of matter stay the same after mixing? In science, a combination of materials is called a **mixture** when the materials do not change into something else after they are mixed. In this investigation, you'll determine whether matter is conserved when baking soda and water form a mixture.

Materials

baking soda	spoon	balance
resealable bag	cup of water	gram masses

NEXT GENERATION SCIENCE STANDARDS | DISCIPLINARY CORE IDEAS
PS1.A: Structure and Properties of Matter
The amount (weight) of matter is conserved when it changes form, even in transitions
in which it seems to vanish. (5-PS1-2)

34

1 Measure a half spoonful of baking soda and put it into the bag. Seal the bag.

2 Place the bag of baking soda and the cup of water on the balance. Record the total mass of the materials.

3 Open the bag. Carefully empty the contents into the cup of water. Use the spoon to stir the mixture until the baking soda is completely dissolved in the water.

4 Place the cup of baking soda water and the empty bag on the balance. Record the total mass of the materials.

A fruit smoothie is a tasty mixture! The same amount of ingredients remain in the blender after they are mixed.

Wrap It Up!

1. **Compare** How did the mass of the materials before the baking soda and water were mixed compare to the mass of the materials after they were mixed?

2. **Analyze** Why was the empty bag added to the balance in step 4?

3. **Draw Conclusions** How do your findings demonstrate the conservation of matter?

Provide Evidence

You've observed that matter is conserved, even after it is changed by cooling, heating, or mixing. Now it's your turn to provide evidence of the conservation of matter. You'll develop an investigation to measure and graph the mass of matter before and after a physical change of your choice. Examples of changes could include heating, cooling, or mixing.

1. **Ask a question.**
   How can you measure and graph quantities to provide evidence that matter is conserved?

2. **Plan and conduct an investigation.**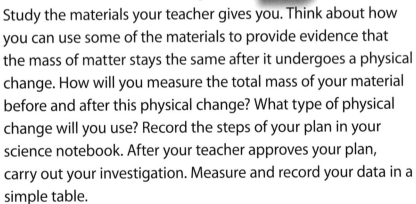
   Study the materials your teacher gives you. Think about how you can use some of the materials to provide evidence that the mass of matter stays the same after it undergoes a physical change. How will you measure the total mass of your material before and after this physical change? What type of physical change will you use? Record the steps of your plan in your science notebook. After your teacher approves your plan, carry out your investigation. Measure and record your data in a simple table.

3. **Analyze and interpret data.**
   Make a bar graph of your data. How do the quantities before and after the change compare? How can you explain your results? How do your findings provide evidence of the conservation of matter?

NEXT GENERATION SCIENCE STANDARDS | PERFORMANCE EXPECTATION
5-PS1-2. Measure and graph quantities to provide evidence that regardless of the type of change that occurs when heating, cooling, or mixing substances, the total weight of matter is conserved.

36

4. **Share your results.**

   Share your conclusions with the class. After hearing from your classmates, make a generalization about the conservation of matter.

This glowing liquid will harden into metal. The mass of the metal will remain the same.

Chemical Changes

Unlike the physical changes you have learned about so far, a **chemical change** causes a material to change into an entirely different material with properties that are different from the original material. The process by which a chemical change occurs is called a **chemical reaction.** In many chemical reactions, the new material cannot be changed back to the original material.

The conservation of matter applies to chemical changes as well as physical changes. The total mass of matter before a chemical reaction is the same as the total mass of matter after the reaction.

Inside a glow stick, a chemical reaction releases energy in the form of light.

NEXT GENERATION SCIENCE STANDARDS | DISCIPLINARY CORE IDEAS
PS1.B: Chemical Reactions
• When two or more different substances are mixed, a new substance with different properties may be formed. (5-PS1-4)
• No matter what reaction or change in properties occurs, the total weight of the substances does not change. (5-PS1-2)

1 A glow stick is a hollow plastic tube containing a liquid and a small glass capsule filled with a second liquid.

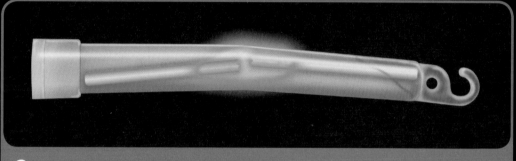

2 When the stick is bent, the capsule inside breaks and the two liquids mix.

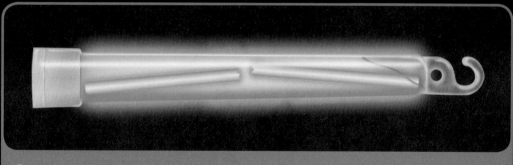

3 A chemical reaction between the two liquids produces light.

My science notebook

Wrap It Up!

1. **Restate** What is a chemical reaction?

2. **Compare and Contrast** How is a chemical change different from a physical change? How is it the same?

Signs of a Chemical Change

Chemical reactions are happening all around you all the time. For instance, chemical reactions digest food in your stomach. A banana that turns brown over time has undergone a chemical reaction. A burning candle is another example of a chemical reaction.

How do you know whether a chemical change has occurred? Learn to spot the signs. Some signs of chemical reactions are bubbles, changes in color, and production of smells, light, and heat.

The smell of burning toast and its dark black color tell you a chemical change has occurred. No matter what reaction or chemical change occurs, the total weight of the substances does not change.

NEXT GENERATION SCIENCE STANDARDS | DISCIPLINARY CORE IDEAS
PS1.B: Chemical Reactions
• When two or more different substances are mixed, a new substance with different properties may be formed. (5-PS1-4)
• No matter what reaction or change in properties occurs, the total weight of the substances does not change. (5-PS1-2)

The smell, the light, and the heat are three signs of chemical change in a burning sparkler.

When a weak acid is dropped on limestone, bubbles are produced. Bubbles are one sign of a chemical change.

My science notebook

Wrap It Up!

1. **Identify** List five signs that a chemical reaction has occurred.

2. **Apply** Identify some chemical reactions that you observe regularly at home.

Chemical Reactions

? **How can you show that a new substance forms when some materials are mixed?**

You can look for signs that chemical reactions are happening. The light and heat of a campfire, for instance, are signs that a chemical change is taking place. You can infer that the wood is changing into a different material. The wood is combining with oxygen from the air and changing into carbon dioxide gas, water, and other materials in the ash and smoke. In this investigation, you'll observe a chemical reaction between an effervescent tablet and water.

Materials

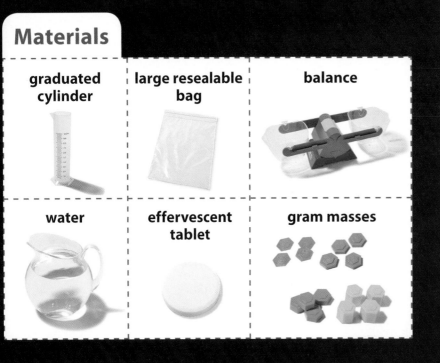

graduated cylinder	large resealable bag	balance
water	effervescent tablet	gram masses

1 Hold the bag open while your partner pours 100 mL of water into the bag. Seal the bag and place it on the balance along with the effervescent tablet. Record the mass of your starting materials.

2 Unzip about 2.5 cm of the zip lip. Prepare yourself to re-zip the bag as soon as your partner adds the tablet. Add the tablet, and immediately re-zip the bag.

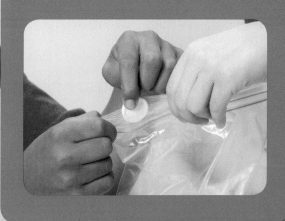

3 Observe what is happening to the contents of the bag. Record your observations.

4 After the tablet stops fizzing, find the mass of the bag and its contents again. Record the mass of your resulting materials.

Wrap It Up!

1. **Describe** What did you observe when you added the tablet to the water?

2. **Infer** What evidence shows that a new substance formed?

3. **Draw Conclusions** How do your findings demonstrate the conservation of matter?

Identify Materials

You've seen that materials can be identified by their properties. You've also seen how materials change as a result of heating and cooling, and what happens in chemical reactions. Now it's your turn. Imagine that you are helping to prepare dinner. To "help," someone has measured out sugar, cornstarch, baking soda, and baking powder. Unfortunately, the powders are not labeled! But these powders have different properties when mixed with liquids such as water, vinegar, and iodine. It will be your job to use the properties of these materials to identify which is which.

1. **Ask a question.**
   How can you use observations and measurements to identify materials based on their properties?

2. **Plan and conduct an investigation.** my science notebook
   Your teacher will provide four powder samples and three liquids. Study the powders and liquids. You should make careful observations, but do not taste any of the materials. Think about how you could test the powders to determine what they are. How will you mix each powder with each liquid? How will you measure the materials? How will you record your observations? Record the steps of your plan in your science notebook. After your teacher approves your plan, carry out your investigation. As you work, record your data in a chart. Include details on the characteristics of each powder and how each powder reacts to each of the liquids.

3. **Analyze your results.**
    After you conduct your tests, compare your
    results to the powder ID chart your teacher
    will provide. Use the chart and your data to
    determine which powder is which. Discuss the
    scientific basis of your findings. What evidence
    helped you to identify each powder?

4. **Share your results.**
    Share your conclusions with the class. As a class,
    discuss the observations that were most useful
    in identifying each powder.

Knowing how materials react enables
firework makers to choose the right
chemicals to burn, explode in certain
ways, and produce colors as planned.

Research Scientist

Genghis Khan was a legendary ruler and warrior. In the early 13th century, he founded the largest empire in the known world. It is thought that he died in 1227 during a military battle, but the details of his death remain a mystery. His tomb has never been found.

That is where Dr. Albert Yu-Min Lin comes in. As a research scientist, he plans and carries out scientific investigations. Albert wants to know what happened to Genghis Khan, and he wants the public to help him. In his work with the Valley of the Khans Project, he uses a technique called crowdsourcing. This means he invites people all over the world to help him search through huge amounts of satellite imagery data for clues.

Albert uses a variety of different technologies in his hunt for the lost tomb. They include satellite imagery, ground-penetrating radar, and even electromagnets. He hopes to one day locate Genghis Khan's burial site and answer the many questions surrounding the ruler's mysterious death.

Albert navigates a computer model of terrain in a virtual reality environment called the StarCAVE.

NEXT GENERATION SCIENCE STANDARDS | CONNECTIONS TO NATURE OF SCIENCE
Scientific Investigations Use a Variety of Methods
Science investigations use a variety of tools and techniques.

Albert Yu-Min Lin is a research scientist at the University of California, San Diego. His quest for Genghis Khan's tomb is featured in a documentary called *The Forbidden Tomb of Genghis Khan* and has taken him to some of the most isolated areas in the world. Albert also enjoys public speaking, mountain climbing, surfing, and photography.

The wall of computer screens behind Albert displays the most detailed satellite images available. The display shows part of Mongolia, a very challenging place to explore.

Life Science

Matter and Energy in Organisms and Ecosystems

The Blue-fronted
Amazon parrot
feeds on fruit in a
tropical rain forest in
South America.

What Plants Need

n the treetops of a rain forest, an
orchid plant clings to the branch of a
huge tree. The orchid and the tree are
very different. Yet both of these plants
need three main things to grow—
sunlight, air, and water.

Like all living things, plants require
food for energy. But unlike animals,
plants are able to make their own food.
They do this by using the energy
of sunlight.

The orchid gets water from rain and fog.

n a rain forest, the tall trees compete for sunlight.
Their leaves block much of the light. Only a few kinds
of plants can grow in the deep shade of the forest
floor. Many small plants, such as this orchid, grow on
the tall trees' branches. By growing on another plant,
the orchid is able to get the sunlight it needs
to survive.

NEXT GENERATION SCIENCE STANDARDS | DISCIPLINARY CORE IDEAS
PS3.D: Energy in Chemical Processes and Everyday Life
The energy released [from] food was once energy from the sun that was captured by
plants in the chemical process that forms plant matter (from air and water). (5-PS3-1)

No soil needed! This orchid is growing on the branches of a tree in a rain forest.

Wrap It Up!

1. **Recall** What are the three main things that plants need to survive?

2. **Identify** What source of energy do plants use to make food?

3. **Summarize** How does an orchid in the rain forest get what it needs to survive?

51

How Plants Get Energy

Plants make their own food in a chemical process called **photosynthesis.** Photosynthesis takes place in leaves, where the green pigment **chlorophyll** captures the energy of sunlight. Leaves use this energy to change carbon dioxide and water into sugar. The carbon dioxide comes from the air. The water is taken up from soil or from the air.

The main product of photosynthesis is food in the form of sugar. Sugar stores energy. Plants use the energy released from sugar to carry out their basic functions. So the energy released from sugar was once energy from the sun. Photosynthesis also produces oxygen, which is released into the air. The diagram shows the chemical process of photosynthesis.

Sunlight provides the energy that plants use to produce their food.

NEXT GENERATION SCIENCE STANDARDS | DISCIPLINARY CORE IDEAS
PS3.D: Energy in Chemical Processes and Everyday Life
The energy released [from] food was once energy from the sun that was captured by plants in the chemical process that forms plant matter (from air and water). (5-PS3-1)

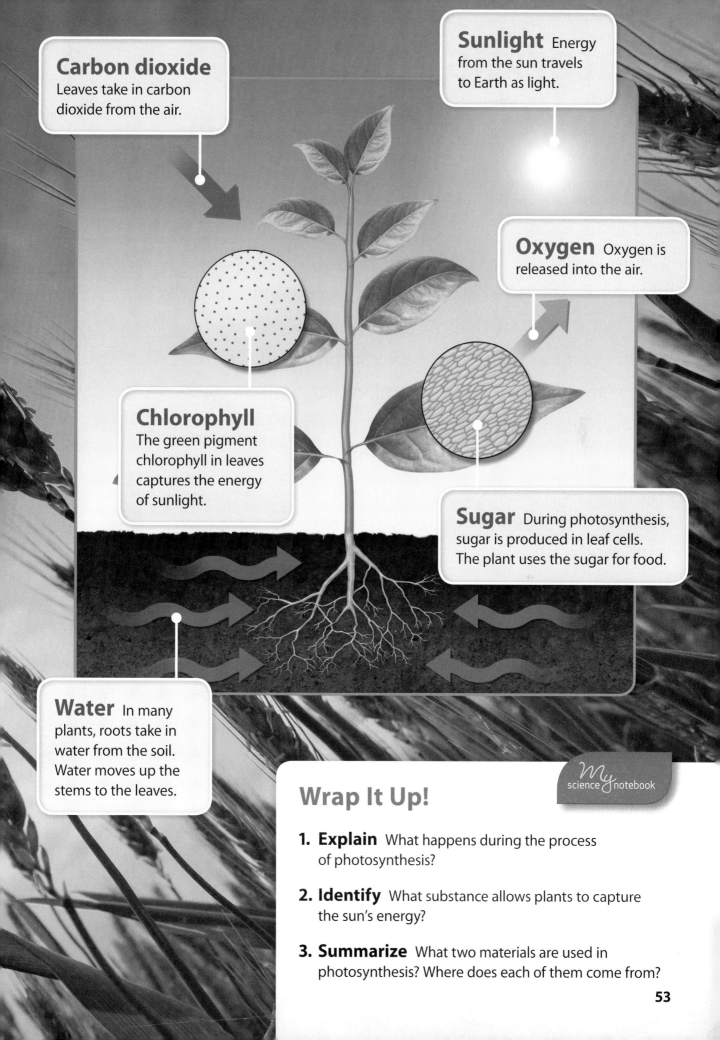

Carbon dioxide
Leaves take in carbon dioxide from the air.

Sunlight Energy from the sun travels to Earth as light.

Oxygen Oxygen is released into the air.

Chlorophyll
The green pigment chlorophyll in leaves captures the energy of sunlight.

Sugar During photosynthesis, sugar is produced in leaf cells. The plant uses the sugar for food.

Water In many plants, roots take in water from the soil. Water moves up the stems to the leaves.

My science notebook

Wrap It Up!

1. **Explain** What happens during the process of photosynthesis?

2. **Identify** What substance allows plants to capture the sun's energy?

3. **Summarize** What two materials are used in photosynthesis? Where does each of them come from?

53

Materials for Plant Growth

Wild bamboo plants grow very quickly. Some can grow more than 60 centimeters (2 feet) in a single day!

Where does the material for this amazing growth come from? You may think it comes from soil, but it does not. Most of it comes from carbon dioxide, a gas in the air, and water. Remember that, during photosynthesis, plants capture energy from the sun and produce sugar. Plants use that sugar as a source of stored energy. Plants also use sugar as a building block to make new leaves, stems, and roots.

In addition to air and water, plants need mineral **nutrients,** which include nitrogen, calcium, and magnesium. Plants use these nutrients to make important substances such as proteins and DNA.

Mineral nutrients are dissolved in water. As a plant takes in water, it also takes in nutrients. When gardeners and farmers add fertilizers to soil or water, they are adding mineral nutrients.

Bamboo grows very quickly. It gets the materials for growth from air, water, and dissolved nutrients.

The nutrients these bamboo plants need are dissolved in the water in the vase. The bamboo does not need soil to live and grow.

Wrap It Up!

1. **Recall** Where does most of the material in a plant come from?

2. **Describe** Where do plants get mineral nutrients?

3. **Analyze** Sometimes people call fertilizers "plant food." Is this an accurate way to describe mineral nutrients? *Hint*: Do mineral nutrients provide a plant with energy?

Growing Crops

Problem

How can people grow crops where there is not enough land with good soil?

Nearly all plants grow in soil. Some places on Earth have thick, fertile soil that supports crop growth. But in many parts of the world, there is a shortage of land with good soil. Even in places where the soil is rich, other problems such as pests and too much or too little rain can harm crops.

Why is crop growth important? Today there are about 7 billion people in the world. By 2050 that number is expected to increase to more than 9 billion. All those people will need much more food. As the population grows, more and more land will be used for cities, roads, and homes. Where will that food be grown?

Insects and other pests may devour crops, leaving little food for humans. Raccoons damaged this sweet corn.

NEXT GENERATION SCIENCE STANDARDS | DISCIPLINARY CORE IDEAS
LS1.C: Organization for Matter and Energy Flow in Organisms
56 Plants acquire their material for growth chiefly from air and water. (5-LS1-1)

In low-lying areas, frequent floods can destroy crops.

Some crops are damaged when rain does not fall regularly. Severe lack of rain is called drought.

In some parts of the world, the soil lacks nutrients or is too dry for crops to grow.

Solution

One solution is hydroponics. In **hydroponics,** plants are grown in water instead of soil. Nutrients plants would normally get from water in soil are added to the water in which the plants grow. Because these systems are often set up in greenhouses, fresh local food is made available year-round.

You might think that hydroponic systems would use more water than traditional methods of farming. But hydroponics can actually save water, since the plants do not have to be irrigated. That makes hydroponics very useful in arid regions. Plants still get sunlight and air they need. In parts of Africa, scientists have found that crops grown with hydroponics use only one third of the water used by a traditional farm.

These vegetables are being grown at the South Pole. The hydroponic grow room produces enough vegetables to provide 50 people at least one salad a week.

Plants in hydroponic systems are often grown in racks so that they take up little space.

Wrap It Up!

1. **Define** What is hydroponics?

2. **Compare and Contrast** Compare the way that plants grown in soil and plants grown by hydroponics get the things they need. Include sunlight, water, air, and mineral nutrients.

3. **Apply** As the world population increases, there may be less land available to grow food. How can hydroponics help solve this problem?

59

Hydroponics

? **How can you grow plants without soil?**

If people travel to Mars, how could they grow food for themselves? There isn't any soil in space. One solution would be to use hydroponics. Such a system could easily produce vegetables, such as tomatoes and lettuce. Imagine eating a fresh salad while traveling to Mars!

Does it seem amazing that astronauts can grow plants without soil? In this investigation, you will gather evidence to show that it is possible.

Materials

clear plastic container and lid with a hole in the middle	young plant	water	liquid houseplant fertilizer
		cotton	

1 Pour water into the container until it is half full. Add 5 drops of liquid houseplant fertilizer to the water to provide nutrients for the plant.

2 Gently push the roots of the plant through the hole in the lid. Then wrap the stem with cotton. Attach the lid to the container.

3 Place the container in a sunny place, such as a windowsill. If necessary, add more water so the roots are covered. Observe and record.

4 Observe the plant for the next few days. Add water to the container to keep the roots underwater. Record your observations.

Wrap It Up!

1. **Observe** How did the plant change during the period you were watching it?

2. **Draw Conclusions** Use evidence from your investigation to support an argument that plants can grow without soil.

3. **Explain** What was the source of the materials that the plant used to grow larger?

4. **Infer** What would happen if you put the plant in a dark closet? Why?

Support an Argument

Orchids, bamboo plants, and corn plants are very different from each other, but they all need to take in materials for growth. Because most of the plants we see grow in soil, many people think plants use soil as a building block for their growth. How could you support an argument that describes where plants get the materials they need to live and grow? Use examples and evidence from pages 54–61 to decide whether plants need the following materials to live and grow:

- water
- soil
- carbon dioxide from air
- nutrients

1. **List.**

   Which materials did you decide that plants need to live and grow? What examples or evidence did you use to make your decision?

2. **Compare.**

   Work with a group. Compare your lists. Discuss your choices, and work together until you can agree about which materials are needed for plants to live and grow.

How are these plants getting the materials they need for growth?

Consumers get energy by eating other organisms. Many consumers eat plants, but other consumers eat animals. Still other consumers eat both plants and animals.

A food chain is a path by which energy flows from one living thing to another in an environment. The photos below show a food chain in a desert. The arrows show one way that energy moves from one living thing to another. Use your finger to trace the path of energy.

Texas horned lizard

Consumer
The Texas horned lizard gets energy by eating the grasshopper.

red-shouldered hawk

Consumer
The hawk eats the lizard. The sun's energy has moved through the food chain.

My science notebook

Wrap It Up!

1. **Define** What is a food chain?

2. **Compare** How do producers and consumers each obtain the energy they need to live and grow?

3. **Analyze** Could producers live without consumers? Could consumers live without producers? Explain.

Compare and Contrast

The diagrams you see here show food chains in two different environments. One food chain is in a pond and the other is in a rain forest.

In any food chain, the energy contained in food was once energy from the sun. Plants capture the sun's energy in chemical processes that form plant matter. Tiny, floating algae are not plants, but like plants, they are producers. They use energy from the sun to produce their own food.

Compare the two food chains shown here. How are they alike? How are they different?

sunlight

Energy Source
Sunlight is the source of energy for the food chain.

sunlight

Energy Source
Sunlight is the source of energy for the food chain.

NEXT GENERATION SCIENCE STANDARDS | DISCIPLINARY CORE IDEAS
PS3.D: Energy in Chemical Processes and Everyday Life
The energy released [from] food was once energy from the sun that was captured by plants in the chemical process that forms plant matter (from air and water). (5-PS3-1)

POND FOOD CHAIN

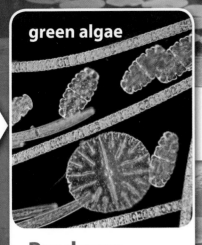

green algae

Producer
Algae use sunlight to grow and reproduce.

water fleas

Consumer
Tiny animals such as water fleas eat algae.

blue gill sunfish

Consumer
Small fish eat smaller animals such as water fleas.

RAIN FOREST FOOD CHAIN

red cocoa fruit

Producer
Rain forest trees use sunlight to make food. They use the energy in this food to produce fruit and seeds.

spider monkey

Consumer
Monkeys eat fruits and seeds.

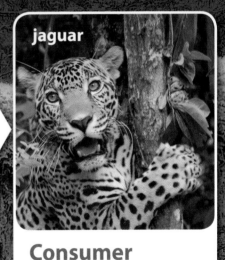

jaguar

Consumer
Jaguars eat animals, such as monkeys.

Wrap It Up!

1. **Interpret Diagrams** What is the original source of energy for both food chains?

2. **Compare and Contrast** How are the producers and consumers in the pond like those in the rain forest? How are they different?

Use Models

You've learned that all animals need energy to grow, move, and repair their bodies. You've also seen some of the ways animals get this energy from their environment through food chains. Now it's your turn. Using your own research, you will make and present a model food chain for an environment. The food chain should include at least two different kinds of animals.

1. **Ask a question.**

   How can you use a model to describe that energy in animals' food was once energy from the sun?

2. **Research an environment.**

   First pick an environment that you would like to learn about. Then use print and online reference materials to discover what kinds of plants and animals live in that environment. Be sure to record in your science notebook the source of all the information you find.

   After you have some information, think about the organisms you have identified. Which are the producers? What consumers eat those producers? And what consumers eat those consumers? Identify one or two pathways that energy takes through the environment.

NEXT GENERATION SCIENCE STANDARDS | PERFORMANCE EXPECTATION
5-PS3-1. Use models to describe that energy in animals' food (used for body repair, growth, motion, and to maintain body warmth) was once energy from the sun.

70

3. **Assemble your model.**
   How will you make your model food chain? You might use index cards, clay models, or computer slide-show presentation software. Include labels for each step in the chain. Your labels should give a brief description of how each organism gets energy.

4. **Analyze and revise your model.**
   Do research to find another plant or animal from your selected environment. Add this organism to the model food chain.

5. **Present your model.**
   When you have revised your model, present your model food chain to the class. Point out how each organism in the food chain gets energy for life processes. Then use your model to explain how all food energy was once energy from the sun.

The sun is the original source of energy in every food chain on the African savanna.

Desert Food Web

A rattlesnake in the desert eats animals such as mice and grasshoppers. These animals are part of the desert **food web.** A food web is a combination of food chains that shows how energy moves from the sun through an environment. Food webs show that consumers get energy from a variety of organisms.

mountain lion

badger

lubber grasshopper

plants

NEXT GENERATION SCIENCE STANDARDS | DISCIPLINARY CORE IDEAS
LS2.A: Interdependent Relationships in Ecosystems
The food of almost any kind of animal can be traced back to plants. Organisms are related in food webs in which some animals eat plants for food and other animals eat the animals that eat plants. Organisms can survive only in environments in which their particular needs are met. A healthy ecosystem is one in which multiple species of different types are each able to meet their needs in a relatively stable web of life. (5-LS2-1)

Trace the Energy! The arrows show the direction in which energy moves through the food web. For example, plants get energy from sunlight. The bighorn sheep gets energy from plants. The cougar gets energy from the bighorn sheep. Use your finger to trace the energy through different food chains in the desert food web.

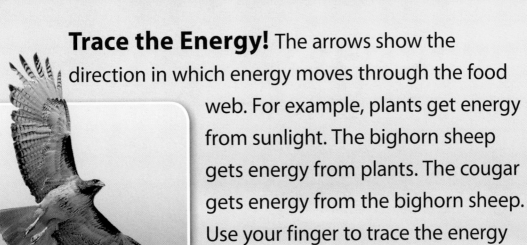

red-tailed hawk

red diamond rattlesnake

bighorn sheep

plants

harvest mouse

Wrap It Up!

My science notebook

1. **Explain** How does energy flow through a food chain?

2. **Contrast** How is a food web different from a food chain?

3. **Infer** Suppose a disease kills most of the hawks in a part of the desert. How might the loss of the hawks affect the other animals in the area?

Decomposers

The mushroom in the photo is neither a producer nor a consumer. Where does it get its energy? The mushroom gets its energy from dead leaves, wood, and other parts of decaying plants. The mushroom is a **decomposer,** an organism that gets its energy by breaking down dead organisms or the waste of living things. **Fungi,** such as the mushroom, and microscopic organisms called **bacteria** are two kinds of decomposers.

When any plant or animal in a food web dies, it is broken down by decomposers. Decomposition eventually restores, or recycles, nutrients back to the soil. The nutrients released by decomposers help plants grow.

These red Amanita mushrooms get energy by breaking down leaves and other plant parts that fall onto the forest floor.

NEXT GENERATION SCIENCE STANDARDS | DISCIPLINARY CORE IDEAS
LS2.A: Interdependent Relationships in Ecosystems
The food of almost any kind of animal can be traced back to plants. Some organisms, such as fungi and bacteria, break down dead organisms (both plants or plant parts and animals) and therefore operate as "decomposers." Decomposition eventually restores (recycles) some materials back to the soil. A healthy ecosystem is one in which multiple species of different types are each able to meet their needs in a relatively stable web of life. (5-LS2-1)

These bacteria are decomposers that get their energy from dead leaves. The image is greatly magnified.

Wrap It Up!

my science notebook

1. **Name** What are two kinds of decomposers?

2. **Explain** How do decomposers get energy?

3. **Cause and Effect** Suppose there were no decomposers in the soil. How might this affect plants growing in the area?

75

Cycles of Matter

The grizzly bear walking among the grasses in the meadow may look like he is alone, but he's not! Other animals such as rabbits, mice, and deer live here. Fungi and bacteria, as well as insects and earthworms, live in the soil. All of these organisms are connected by cycles of matter. These cycles also include the air and soil.

All of the organisms in the meadow play a part in the carbon dioxide-oxygen cycle.

Carbon Dioxide The grizzly bear breathes in oxygen. It uses the oxygen to break down and get energy from food. The bear gives off carbon dioxide as waste. Plants use carbon dioxide to make food.

NEXT GENERATION SCIENCE STANDARDS | DISCIPLINARY CORE IDEAS
LS2.B: Cycles of Matter and Energy Transfer in Ecosystems
Matter cycles between the air and soil and among plants, animals, and microbes as these organisms live and die. Organisms obtain gases, and water, from the environment, and release waste matter (gas, liquid, or solid) back into the environment. (5-LS2-1)

Carbon dioxide and oxygen are gases found in the air. The carbon dioxide-oxygen cycle provides living things with the carbon and oxygen they need to survive.

Organisms in the meadow are also connected by the nitrogen cycle. Nitrogen is a substance found in the air and in the soil. Plants take in nitrogen from the soil. Animals get the nitrogen they need by eating plants or other animals. When plants and animals die, decomposers return the nitrogen in their bodies to the soil. Plant and animal wastes also contain nitrogen, which microbes return to the soil.

Oxygen When plants make their own food through photosynthesis, they take in carbon dioxide from the air and give off oxygen as waste. They also use oxygen to break down and get energy from food.

Wrap It Up!

My science notebook

1. **Describe** What is the role of decomposers in the nitrogen cycle?

2. **Explain** Why is the carbon dioxide-oxygen cycle important to plants and animals?

3. **Sequence** The following organisms are part of the nitrogen cycle: microscopic decomposers, plant, rabbit. Draw a diagram with arrows that puts the organisms in the correct order. Begin with nitrogen in the soil.

Tallgrass Prairie Ecosystem

These bison live in the tallgrass prairie of North America. All of the plants and animals in a tallgrass prairie interact with the nonliving things in their environment. Some of these things are physical characteristics, such as the summer and winter temperatures, amount of rainfall, and kind of soil. Grasses grow in the deep, fertile soil. Animals drink water and breathe the air. These organisms also interact with the many other living things in the prairie.

These bison live in the Tallgrass Prairie Reserve in Oklahoma.

NEXT GENERATION SCIENCE STANDARDS | DISCIPLINARY CORE IDEAS
LS2.A: Interdependent Relationships in Ecosystems
The food of almost any kind of animal can be traced back to plants. Organisms are related in food webs in which some animals eat plants for food and other animals eat the animals that eat plants. Organisms can survive only in environments in which their particular needs are met. A healthy ecosystem is one in which multiple species of different types are each able to meet their needs in a relatively stable web of life. (5-LS2-1)

You would not find wild bison living in a desert. The grasses the bison eat do not grow in the dry conditions of a desert. Organisms that live in the tallgrass prairie are able to get what they need to survive from this ecosystem. An **ecosystem** is all the living and nonliving things that interact in an area. A healthy ecosystem is one in which many types of living things are able to meet their needs.

Horned larks live in the tallgrass prairie. They weave their nests from the fine grasses. Adults feed on the seeds of grasses and wildflowers. They feed insects to their young.

These burrowing owls live in a nest dug in the soil of the prairie.

My science notebook

Wrap It Up!

1. **Define** What is an ecosystem?

2. **Infer** What are some of the nonliving things you can observe or infer in this photo of a tallgrass prairie?

3. **Explain** How do the physical characteristics of an environment help support the organisms that live there?

Grassland Populations and Communities

Many different kinds of organisms, or **species,** live in the prairie ecosystem. Scientists classify the organisms in an ecosystem into three levels—individual organisms, populations, and communities. Individual prairie dogs usually live together in prairie dog towns. A **population** is all the individuals of a species living together in a particular place.

Individual
A single organism, such as this prairie dog, is an individual in an ecosystem.

Population
All the prairie dogs that live in a particular part of the ecosystem are a population.

NEXT GENERATION SCIENCE STANDARDS | DISCIPLINARY CORE IDEAS
LS2.A: Interdependent Relationships in Ecosystems
The food of almost any kind of animal can be traced back to plants. Organisms are related in food webs in which some animals eat plants for food and other animals eat the animals that eat plants. Organisms can survive only in environments in which their particular needs are met. A healthy ecosystem is one in which multiple species of different types are each able to meet their needs in a relatively stable web of life. (5-LS2-1)

Prairie dogs live in burrows, which they dig in the ground. Other animals, such as burrowing owls and black-footed ferrets, share their burrows. Prairie dogs eat the grasses and wildflowers that live in the prairie. They also serve as food for other animals, such as black-footed ferrets, foxes, and hawks. All of the populations that live and interact in an area make up a **community.** Organisms in a community can only survive in environments in which their particular needs are met. Healthy communities have many species connected by a variety of food webs.

Community

All the populations of organisms that live and interact in that part of the prairie form a community.

Ecosystem

All of the communities plus the physical parts of the environment that interact together make up the ecosystem.

My science notebook

Wrap It Up!

1. **List** What are the three levels of organisms that make up an ecosystem?

2. **Compare** How is a population different from a community?

Interactions in a Model Pond

? **How do living things in a model pond ecosystem interact?**

An ecosystem is all the living and nonliving elements in an area. You can make a model pond ecosystem and explore some of the interactions that take place there.

Materials

clear plastic bottle	sand	small rocks	Elodea
water	spoon	3 snails	hand lens

NEXT GENERATION SCIENCE STANDARDS | DISCIPLINARY CORE IDEAS
PS3.D: Energy in Chemical Processes and Everyday Life
The energy released [from] food was once energy from the sun that was captured by plants in the chemical process that forms plant matter (from air and water). (5-PS3-1)

1 Layer sand and rocks at the bottom of a clear plastic bottle. Plant the Elodea. Pour water into the bottle until it is about two-thirds full. Use a spoon to place 3 snails in the model ecosystem.

2 Put your model in a sunny place. Observe the model each day for 7 days. Record your observations in your science notebook. Use a hand lens to observe changes in the ecosystem.

3 Use your observations to infer what each living thing needs and how it meets those needs. Classify each organism as a producer or consumer.

4 Draw your model pond ecosystem. Label the organisms. Draw arrows to show how energy moved from the sun to the producers to the consumers in the model. Then use the Internet or other resources to research one more producer and one more consumer that could be added to your model pond ecosystem. Revise your drawings of the model ecosystem based on the addition of these organisms.

Wrap It Up!

1. **Classify** How did your observations help you classify producers and consumers in your ecosystem?

2. **Compare and Contrast** In what ways is your model like a real pond? In what ways is it different?

Develop a Model

Ecosystems are made up of the organisms and nonliving things in the environment. Materials are constantly moving among these different parts. You can develop a model to describe this movement of matter.

1. **Ask a question.**
   How can you develop a model to describe the movement of matter among plants, animals, decomposers, and the environment?

2. **Research an ecosystem.**
   Select an ecosystem for your model. Will it be a small ecosystem, such as a pond, or a big ecosystem, such as a pine forest? Use reliable reference materials to discover how materials cycle through that ecosystem.

   Decide which cycle to show. Where do the raw materials for the cycle come from? Are they in the air, water, or soil? How are the materials taken in by living things? How do the materials move among different organisms? What happens to the materials when organisms die? Record all your findings in your science notebook.

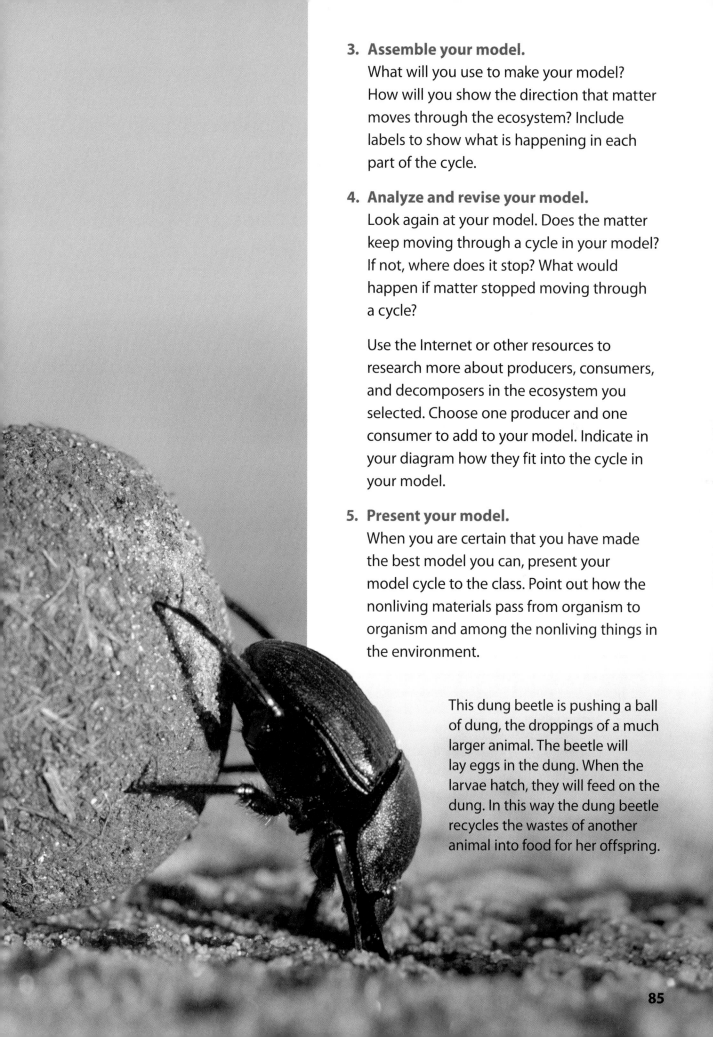

3. **Assemble your model.**
   What will you use to make your model? How will you show the direction that matter moves through the ecosystem? Include labels to show what is happening in each part of the cycle.

4. **Analyze and revise your model.**
   Look again at your model. Does the matter keep moving through a cycle in your model? If not, where does it stop? What would happen if matter stopped moving through a cycle?

   Use the Internet or other resources to research more about producers, consumers, and decomposers in the ecosystem you selected. Choose one producer and one consumer to add to your model. Indicate in your diagram how they fit into the cycle in your model.

5. **Present your model.**
   When you are certain that you have made the best model you can, present your model cycle to the class. Point out how the nonliving materials pass from organism to organism and among the nonliving things in the environment.

This dung beetle is pushing a ball of dung, the droppings of a much larger animal. The beetle will lay eggs in the dung. When the larvae hatch, they will feed on the dung. In this way the dung beetle recycles the wastes of another animal into food for her offspring.

Plants Invade!

Sometimes a newly introduced species can damage the balance of an ecosystem. Kudzu is a plant that is native to Japan and southern China. People brought kudzu to the United States in the late 1800s. Gardeners planted it because of its pretty vines and sweet smell. Farmers planted it to help control erosion.

Kudzu is nicknamed "the vine that ate the South."

But after kudzu was introduced, it began to crowd out native trees and shrubs. It has now spread throughout much of the southeastern United States. Kudzu is an invasive species. An **invasive species** is an organism that does not belong in a certain place and harms the environment. Kudzu grows quickly, covering other plants, utility poles and power lines, and even buildings.

Kudzu distribution, 2009

0   250   500 Miles

0   250 500 Kilometers

This map shows counties in U.S. states where kudzu has invaded. In spite of efforts to stop its spread, kudzu covers 150,000 acres of new land each year.

My science notebook

Wrap It Up!

1. **Define** What is an invasive species?

2. **Infer** How might daily life in the South be affected by the growth of kudzu?

3. **Infer** What could happen if you planted an invasive plant near your home?

Animals Invade!

Problem

Like invasive plants, invasive animals can upset the balance of an ecosystem. The red imported fire ant is an invasive animal that is causing problems in the United States. These ants have a painful sting. They can kill young birds, mammals, and reptiles. They damage crops and can harm or kill young farm animals.

Red imported fire ants have been invading more and more areas of the United States. They don't have any natural enemies in the U.S. How can this invasive species be controlled?

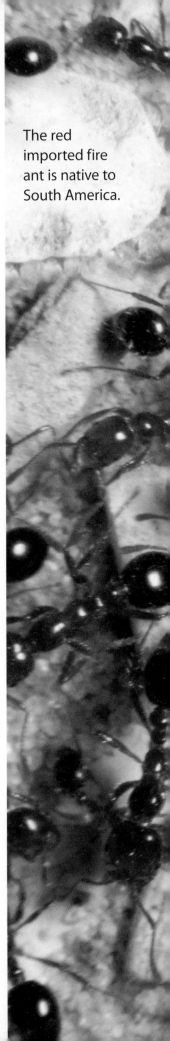

The red imported fire ant is native to South America.

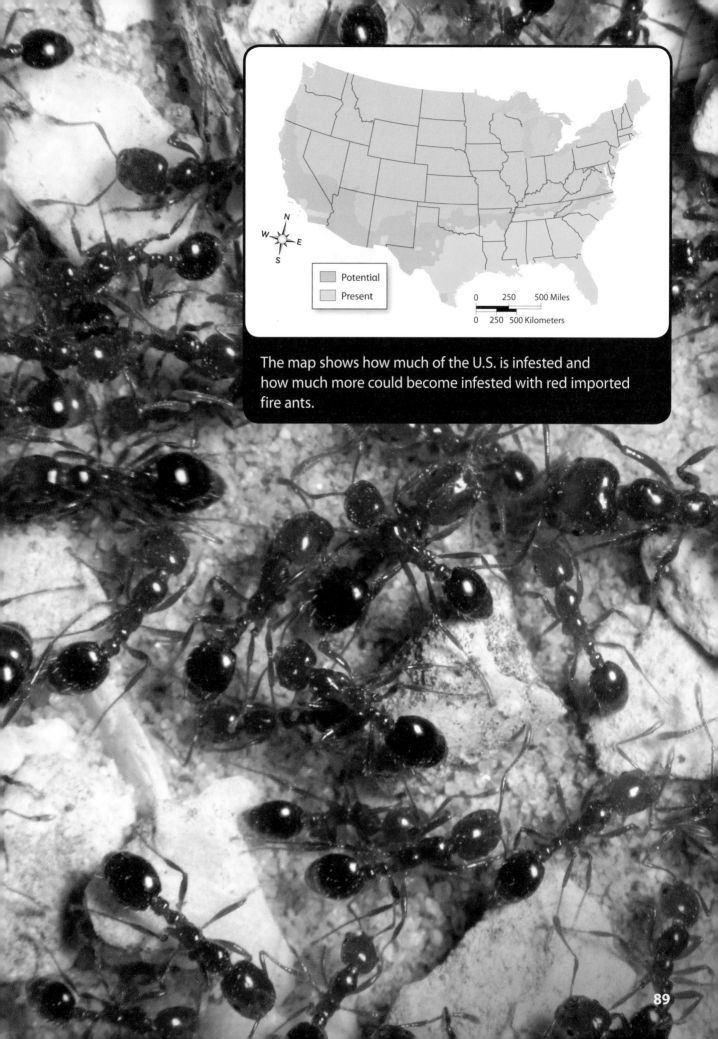

The map shows how much of the U.S. is infested and how much more could become infested with red imported fire ants.

Solution

Recently scientists have been raising and releasing phorid flies in regions that have been invaded by red imported fire ants. Phorid flies are from South America, where they are a natural enemy of red fire ants.

When phorid flies come near red fire ants, the ants try to hide. The flies keep the ants from searching for food. With less food, the fire ant population cannot grow. But why do fire ants hide from phorid flies? It seems that around phorid flies, the ants just can't keep their heads on! Read the captions on the next page to find out why.

phorid fly

red imported fire ant

1 A phorid fly lays an egg in the body of an ant. The egg hatches and the larva moves into the ant's head. It begins to eat the brain. Like a zombie, the ant wanders away from the nest.

2 The larva releases a chemical that causes the ant's head to fall off. The larva continues to feed and develops into a pupa.

3 Finally the adult phorid fly emerges from the ant's head.

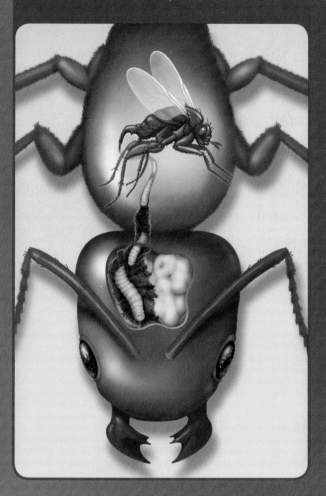

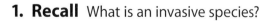

Wrap It Up!

My science notebook

1. **Recall** What is an invasive species?

2. **Cause and Effect** How do phorid flies affect the behavior of red fire ants? How does this affect the red imported fire ant population?

3. **Evaluate** Are phorid flies an invasive species? Explain.

Conservationist

The Colorado River is the largest river in the Southwest. It flows through many states before reaching its mouth, or **delta,** in Mexico. Along the way, people use its water for farms, lawns, cities, and factories. Over the last 100 years, people have been taking more and more water from the river. Now people take so much water that by the time the Colorado reaches the sea, it is only a trickle. There isn't enough water left for the wildlife that live in the delta.

Dr. Osvel Hinojosa is a conservationist who is working with many other people to restore the river's flow. He is also working to restore the wildlife of the delta. For example, he has helped replace invasive saltcedar trees with native trees, such as willows, cottonwood, and mesquite.

NGL Science Why are the wetlands of the delta important?

Dr. Hinojosa Wetlands have a very important function in the world. Many species of plants and animals live in wetlands. Wetlands also help make the water cleaner and provide protection against floods, storms, and hurricanes.

NEXT GENERATION SCIENCE STANDARDS | CONNECTIONS TO NATURE OF SCIENCE
Science Addresses Questions About the Natural and Material World
92 Scientists study the natural and material world. (2-ESS2-1)

Earth's Major Systems

Scientists who study Earth divide it into four major systems. These systems, or spheres, are the geosphere, hydrosphere, atmosphere, and biosphere. Each of these systems is made up

Geosphere

The **geosphere** contains all of Earth's solid and molten, or liquid, rocks. It also includes sediments and soil.

Hydrosphere

The **hydrosphere** is made up of all of the liquid water on Earth, such as the water in rivers, lakes, and the ocean, as well as water found underground. It also includes all of the ice and snow on Earth.

NEXT GENERATION SCIENCE STANDARDS | DISCIPLINARY CORE IDEAS
ESS2.A: Earth Materials and Systems
Earth's major systems are the geosphere (solid and molten rock, soil, and sediments), the hydrosphere (water and ice), the atmosphere (air), and the biosphere (living things, including humans). These systems interact in multiple ways to affect Earth's surface materials and processes. (5-ESS2-1)

of many parts that work together as a whole. The four systems interact in many different ways. Their interactions are constantly changing Earth's surface materials and processes. Study the photos to learn more about these four systems.

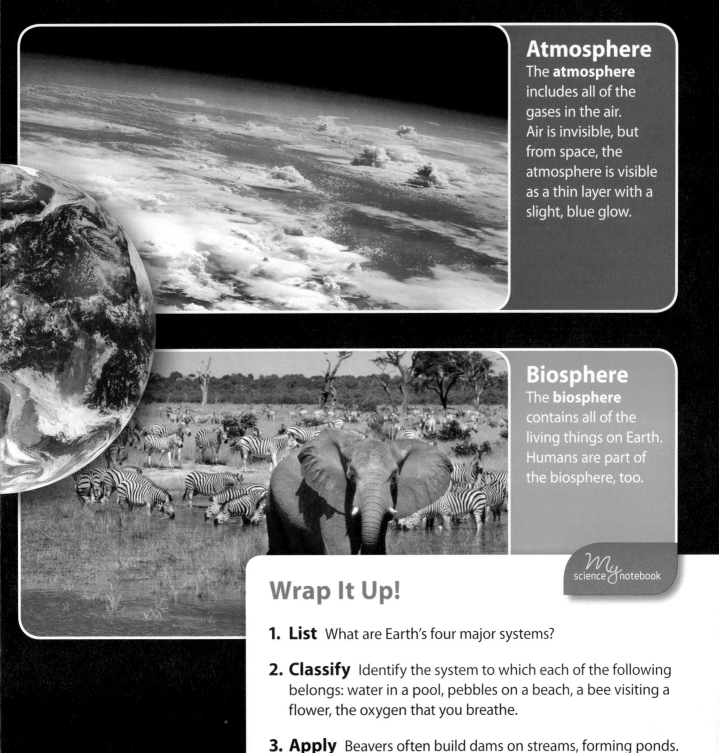

Atmosphere

The **atmosphere** includes all of the gases in the air. Air is invisible, but from space, the atmosphere is visible as a thin layer with a slight, blue glow.

Biosphere

The **biosphere** contains all of the living things on Earth. Humans are part of the biosphere, too.

My science notebook

Wrap It Up!

1. **List** What are Earth's four major systems?

2. **Classify** Identify the system to which each of the following belongs: water in a pool, pebbles on a beach, a bee visiting a flower, the oxygen that you breathe.

3. **Apply** Beavers often build dams on streams, forming ponds. Which systems are interacting when a beaver builds a dam?

The Geosphere

The geosphere is made up of all of the rocks and minerals on Earth's surface. It includes all of the materials in mountain ranges, canyons, beaches, and other landforms. Sediments and soil are a part of the geosphere. The solid rocks and molten materials beneath the surface are also a part of the geosphere.

GEOSPHERE ATMOSPHERE

HYDROSPHERE BIOSPHERE

The geosphere is made up of all of the solid and liquid rock on Earth.

Volcanoes release lava, ash, and gases. These materials change the shape of the land and the composition of the atmosphere.

NEXT GENERATION SCIENCE STANDARDS | DISCIPLINARY CORE IDEAS
ESS2.A: Earth Materials and Systems
Earth's major systems are the geosphere (solid and molten rock, soil, and sediments), the hydrosphere (water and ice), the atmosphere (air), and the biosphere (living things, including humans). These systems interact in multiple ways to affect Earth's surface materials and processes. (5-ESS2-1)

Processes and events in the geosphere are constantly changing the shape of the land. For example, earthquakes crack and move the land. The geosphere also interacts with other systems. Over time, flowing water of the hydrosphere weathers rocks into small pieces. The water erodes the pieces and deposits them in other places. Plants in the biosphere grow in the sediments and soil.

Sediment is made as wind, water, and ice break or weather larger rocks into smaller pieces.

My science notebook

Wrap It Up!

1. **Identify** What are two events or processes that take place in the geosphere?

2. **Cause and Effect** How can water in the hydrosphere change the geosphere?

3. **Explain** A volcano releases large amounts of ash into the air. How might this process affect plants and animals in the biosphere?

The Hydrosphere

GEOSPHERE ATMOSPHERE

HYDROSPHERE BIOSPHERE

The hydrosphere is made up of all of the liquid water on Earth, as well as all of the water frozen in ice and snow. It includes the fresh water in streams, rivers, ponds, lakes, and wetlands. It includes the salt water in the ocean. The clouds in the sky, which are made up of water droplets or particles of ice, are also part of the hydrosphere. **Groundwater**—the water in soil and between the rocks below Earth's surface—is a part of the hydrosphere, too.

The water in the hydrosphere is constantly moving. Rain that falls on land enters rivers, which flow toward the ocean. Water

Clouds Clouds are made up of tiny water droplets or ice crystals.

NEXT GENERATION SCIENCE STANDARDS | DISCIPLINARY CORE IDEAS
ESS2.A: Earth Materials and Systems
Earth's major systems are the geosphere (solid and molten rock, soil, and sediments), the hydrosphere (water and ice), the atmosphere (air), and the biosphere (living things, including humans). These systems interact in multiple ways to affect Earth's surface materials and processes. The ocean supports a variety of ecosystems and organisms, shapes landforms, and influences climate. Winds and clouds in the atmosphere interact with the landforms to determine patterns of weather. (5-ESS2-1)

enters the atmosphere when it **evaporates,** or changes from a liquid to a gas. Eventually water in the atmosphere **condenses** onto dust and other tiny particles in the air, forming clouds. Liquid or solid water falls to the ground as rain or snow.

All organisms require water to survive. A great variety of organisms live in the ocean, while others live in fresh water. On land, plants take in water from the soil. All animals need water to survive. Most land animals cannot survive without taking in water.

Icebergs Icebergs are made up of the solid state of water—ice! Similar to the way ice floats in your lemonade, icebergs float in ocean water.

Salt water The salt water in the ocean is part of the hydrosphere. Most of the water on Earth is found in the ocean. The ocean supports a variety of ecosystems.

My science notebook

Wrap It Up!

1. **Define** What is groundwater?

2. **Identify** What process moves water from Earth's surface to the atmosphere?

3. **Explain** How does water in the atmosphere return to Earth's surface?

The Atmosphere

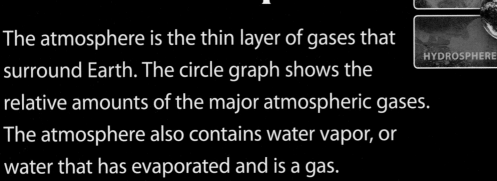

The atmosphere is the thin layer of gases that surround Earth. The circle graph shows the relative amounts of the major atmospheric gases. The atmosphere also contains water vapor, or water that has evaporated and is a gas.

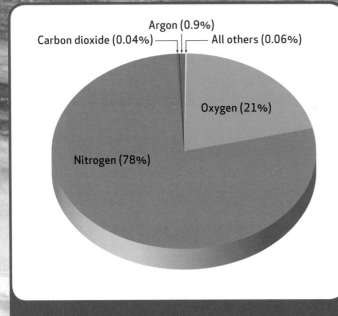

Carbon dioxide (0.04%) — Argon (0.9%) — All others (0.06%)

Oxygen (21%)

Nitrogen (78%)

The most common gases in the atmosphere are nitrogen, oxygen, argon, and carbon dioxide.

NEXT GENERATION SCIENCE STANDARDS | DISCIPLINARY CORE IDEAS
ESS2.A: Earth Materials and Systems
Earth's major systems are the geosphere (solid and molten rock, soil, and sediments), the hydrosphere (water and ice), the atmosphere (air), and the biosphere (living things, including humans). These systems interact in multiple ways to affect Earth's surface materials and processes. The ocean supports a variety of ecosystems and organisms, shapes landforms, and influences climate. Winds and clouds in the atmosphere interact with the landforms to determine patterns of weather. (5-ESS2-1)

The gases in the atmosphere capture some of the energy of sunlight. They also trap heat given off by Earth's surface. This keeps the planet warm. Certain gases protect living things from the sun's harmful rays.

The processes of weather, such as winds and storms, take place in the atmosphere. Winds move sediments across Earth's surface, forming sand dunes and other landforms. Winds blow across lakes and the ocean, forming waves that crash onto the land. Storms produce rain and snow, filling rivers and lakes and changing the surface of the land.

Weather The processes of weather, such as the formation of clouds and storms, take place in the atmosphere.

Wrap It Up!

1. **Interpreting Graphs** Which gas makes up the largest portion of the atmosphere?

2. **Explain** How does wind affect landforms? Give an example.

3. **Infer** The amount of water in the atmosphere differs from place to place. Would you usually expect to find more water in the atmosphere over the land or over the ocean? Explain.

The Biosphere

GEOSPHERE   ATMOSPHERE

HYDROSPHERE   **BIOSPHERE**

The biosphere is made up of all of the organisms living on Earth. It includes all of the plants, animals, fungi, and microbes living. The biosphere also includes human beings.

All living things require water and nutrients to live. Organisms that live in streams, lakes, or the ocean get water from their surroundings. On land, plants get water and nutrients from the soil. Most land animals get water by drinking.

All organisms need water from the hydrosphere.

NEXT GENERATION SCIENCE STANDARDS | DISCIPLINARY CORE IDEAS
ESS2.A: Earth Materials and Systems
Earth's major systems are the geosphere (solid and molten rock, soil, and sediments), the hydrosphere (water and ice), the atmosphere (air), and the biosphere (living things, including humans). These systems interact in multiple ways to affect Earth's surface materials and processes. The ocean supports a variety of ecosystems and organisms, shapes landforms, and influences climate. (5-ESS2-1)

Plants take in carbon dioxide from the atmosphere to make food. They release oxygen into the atmosphere. Animals and plants use oxygen and release carbon dioxide when they break down food to release energy.

Humans are part of the biosphere. So are the crops they grow and living organisms in the soil.

Wrap It Up!

My science notebook

1. **Describe** What makes up the biosphere?

2. **Relate** How do land animals interact with the atmosphere?

3. **Explain** How do elements of the biosphere and the hydrosphere interact?

Earth's Systems Interact

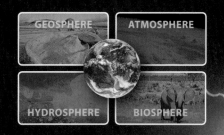

GEOSPHERE ATMOSPHERE

HYDROSPHERE BIOSPHERE

The monsoons of Southeast Asia and India have a powerful effect on the land, ecosystems, and people of the region. A **monsoon** is a strong wind that changes direction with the seasons. These seasonal changes cause the wet and dry seasons of Asia.

Monsoons result from the unequal heating of the land and the ocean. In winter the land is colder than the water. Dry monsoons blow from the land toward the ocean. Plants wither and wells run dry. In summer the water is cooler than the land. The monsoons blow from the ocean toward the land. They bring humid air and large amounts of rain. The rain causes plants to grow rapidly. Humans rely on the summer monsoons to provide water for rice and other crops.

The heavy rains of the summer monsoons may cause damage. Rushing streams sweep away soil. Low-lying areas flood, destroying homes and crops. People and livestock are at risk of drowning.

NEXT GENERATION SCIENCE STANDARDS | DISCIPLINARY CORE IDEAS
ESS2.A: Earth Materials and Systems
Earth's major systems are the geosphere (solid and molten rock, soil, and sediments), the hydrosphere (water and ice), the atmosphere (air), and the biosphere (living things, including humans). These systems interact in multiple ways to affect Earth's surface materials and processes. The ocean supports a variety of ecosystems and organisms, shapes landforms, and influences climate. Winds and clouds in the atmosphere interact with the landforms to determine patterns of weather. (5-ESS2-1)

Summer monsoons can dump as much as 30 centimeters (about 1 foot) of rain in a day!

The heavy rains of the summer monsoons flood low-lying fields.

Water that fell during the summer monsoons permits people to grow crops, such as rice, during the drier winter season.

Wrap It Up!

1. **Define** What is a monsoon?

2. **Compare and Contrast** How are the winter and summer monsoons alike and different? What causes them? Which bring rain and which bring dry air?

3. **Cause and Effect** How do the summer monsoons of India affect the hydrosphere, geosphere, and biosphere?

Interactions of Earth's Systems

? **How can you model interaction among Earth's major systems?**

Earth's four spheres, or systems, interact in multiple ways. You can make a terrarium to model how the geosphere, atmosphere, hydrosphere, and biosphere interact.

Materials

clear plastic bottle with top cut off 	**gravel** 	**potting soil** 
plastic spoon	**small plants** 	**water** **masking tape**

NEXT GENERATION SCIENCE STANDARDS | DISCIPLINARY CORE IDEAS
ESS2.A: Earth Materials and Systems
Earth's major systems are the geosphere (solid and molten rock, soil, and sediments), the hydrosphere (water and ice), the atmosphere (air), and the biosphere (living things, including humans). These systems interact in multiple ways to affect Earth's surface materials and processes. (5-ESS2-1)

In many coastal areas, waves crash onto the shore, wearing down cliffs and breaking rocks apart. The force of the waves may carve rocks into strange shapes. Waves and the sediments they carry eventually grind rocks into tiny bits of sand.

In the process of **erosion,** waves and currents pick up sediments and move them away. In the process of **deposition,** sand and other sediments are laid down in a new place. Deposition builds up on sandy islands and beaches.

Corals formed this barrier reef in Tahiti. The white sands on the reef comes from coral sediments that were deposited by ocean waves.

My science notebook

Wrap It Up!

1. **List** What are two forms of ocean movement that can change the shape of the land?

2. **Contrast** What is the difference between erosion and deposition?

3. **Analyze** The strong winds of large storms increase the size of ocean waves. How might these larger waves affect the amount of erosion on a sandy beach?

115

The Ocean Affects Climate

The ocean has a strong influence on the weather and climate of coastal regions. **Weather** is the state of the atmosphere at a certain place and time. **Climate** is the pattern of the weather in an area over a long period of time.

Places near the ocean usually receive more rainfall than inland areas. The ocean also reduces the extremes of high and low temperatures. This is because water warms and cools more slowly than land.

Frequently, rain falls as moist air above the ocean moves over cooler land.

On a summer day, sunlight heats the land faster than it heats the ocean. Cool breezes blow from the ocean toward the shore. This makes the climate along the coast cooler than the climate farther inland in summer. In winter, the land cools faster than the ocean. The warmer ocean water makes the climate along the coast warmer than that of inland areas in winter.

Ocean currents also influence climate. For example, the Gulf Stream is a current that carries warm water from the Caribbean and Gulf of Mexico into the North Atlantic. Its warm water makes the climate of the East Coast of the United States and the western regions of Europe warmer than in the middle of the continents.

NEXT GENERATION SCIENCE STANDARDS | DISCIPLINARY CORE IDEAS
ESS2.A: Earth Materials and Systems
The ocean supports a variety of ecosystems and organisms, shapes landforms, and influences climate. Winds and clouds in the atmosphere interact with the landforms to determine patterns of weather. (5-ESS2-1)

In this image, reds and oranges indicate warmer temperatures. Greens and blues show cooler temperatures.

The white outline shows the East Coast of North America. The Gulf Stream warms the climate along the coastline.

Waters from the Gulf Stream continue across the Atlantic Ocean. They make the climate in Ireland and western England , which lie farther north than the U.S. East Coast, much warmer than in other places located so far north.

The Gulf Stream carries warm water from the Gulf of Mexico and Caribbean northward.

My science notebook

Wrap It Up!

1. **Define** What is climate?

2. **Cause and Effect** Explain how the Gulf Stream affects the climate of the East Coast of North America.

3. **Summarize** In general, how does the ocean affect the temperature of coastal regions? Explain why.

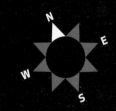

Landforms and Weather Patterns

Weather changes the shape of the land. At the same time, landforms interact with winds and clouds to determine the patterns of weather in a region.

Mountain ranges affect the amount of rain that falls in an area. Winds blowing off the ocean bring moist air to the land. If there is a mountain range in the path of the wind, the air is pushed upward. As the air moves up, it cools. This cooling causes water vapor in the air to condense. Clouds form. Rain and snow from the clouds fall on the side of the mountain closest to the ocean. The air loses much of its moisture.

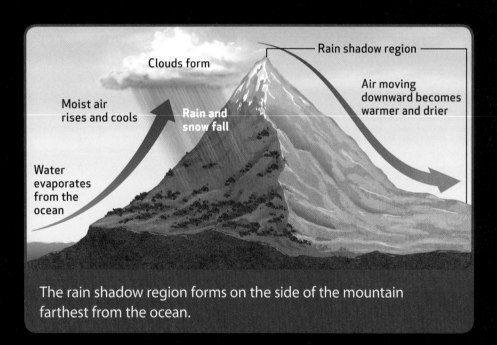

The rain shadow region forms on the side of the mountain farthest from the ocean.

NEXT GENERATION SCIENCE STANDARDS | DISCIPLINARY CORE IDEAS
ESS2.A: Earth Materials and Systems
Earth's major systems are the geosphere (solid and molten rock, soil, and sediments), the hydrosphere (water and ice), the atmosphere (air), and the biosphere (living things, including humans). These systems interact in multiple ways to affect Earth's surface materials and processes. The ocean supports a variety of ecosystems and organisms, shapes landforms, and influences climate. Winds and clouds in the atmosphere interact with the landforms to determine patterns of weather. (5-ESS2-1)

As the air moves down the other side of the mountain range, it gets warmer. Clouds disappear. The area on the dry side of the mountain range is said to be in a **rain shadow.** Little rain falls in these areas.

This satellite image shows a rain shadow on Mauna Loa, a volcano on the island of Hawai'i. Lush greenery grows where warm, moist air rises. Clouds form and rain falls on the coastal side of the volcano.

My science notebook

Wrap It Up!

1. **Explain** Why do clouds form near the top of a mountain range?

2. **Contrast** Describe the difference between the amount of rain that falls on the ocean side of a mountain range and the amount of rain that falls in a mountain range's rain shadow.

3. **Apply** Death Valley in California and Nevada is one of the driest places in North America. Death Valley is located east of the Sierra Nevada. Why is Death Valley so dry? (*Hint*: The winds in this region blow east from the Pacific Ocean.)

The Atmosphere and Landforms

What are these strange-looking rock spires? These formations are tufa towers in Mono Lake, California. Tufa towers contain calcium carbonate, a chemical found in limestone.

Tufa towers form under water. If the water level in the lake drops, the tufa towers are exposed to the atmosphere. Then winds and rain begin to weather the towers. Strong winds that carry sediments act like sandpaper, grinding down the surface of the rocks until they eventually become smooth.

Over time, rain will dissolve the rock until these tufa towers are no longer exposed above the lake's surface.

NEXT GENERATION SCIENCE STANDARDS | DISCIPLINARY CORE IDEAS
ESS2.A: Earth Materials and Systems
Earth's major systems are the geosphere (solid and molten rock, soil, and sediments), the hydrosphere (water and ice), the atmosphere (air), and the biosphere (living things, including humans). These systems interact in multiple ways to affect Earth's surface

Rain weathers rocks and sediments, eventually wearing down hills and even tall mountains. Rain is naturally acidic. Chemical pollution can make rain even more acidic. The acid in rain can dissolve the limestone rocks that make up the towers. The dissolved calcium carbonate runs back into the lake.

My science notebook

Wrap It Up!

1. **List** What are two factors in the atmosphere that can change the shape of landforms?

2. **Cause and Effect** How do the acids in rain affect tufa towers?

3. **Infer** If the amount of acid in rain increases, how might the organisms living in the lake be affected?

Develop a Model

You've learned about Earth's geosphere, hydrosphere, atmosphere, and biosphere and have seen ways in which these systems interact. Now it's your turn. How can you develop a model to describe an interaction between two of Earth's spheres?

1. **Construct an explanatory model.**
   Decide which interaction you will show. Review pages 108–121 to get some ideas. You have many choices. For example, you might show how the ocean affects the shape of the land. You might show how weather and climate affect landforms, or how mountain ranges affect clouds and rain. Be sure the interaction you choose involves two *different* spheres. Draw diagrams that show the interaction.

2. **Design your model.**
   What will you use to make your model? You might make a large poster, a three-dimensional model, or even a computer animation. If your interaction causes a system to change, you could show it as a series of steps.

   Gather your materials and construct your model. Include labels that describe what is happening in each part of the interaction.

3. **Analyze and revise your model.**
   Add a third sphere and explain how it interacts with the other two spheres. Revise your models.

4. **Present your model.**
   When you are sure that your model is the best it can be, present it to the class. In your presentation, identify the spheres involved and describe their interactions.

NEXT GENERATION SCIENCE STANDARDS | PERFORMANCE EXPECTATION
5-ESS2-1. Develop a model using an example to describe ways the geosphere, biosphere, hydrosphere, and/or atmosphere interact.

All of Earth's major systems interact in this coral reef.

Water on Earth

When viewed from space, our planet looks blue. That's because most of Earth's surface is covered with water. Nearly all of Earth's water is in the ocean. Ocean water is salty. Most of Earth's fresh water is frozen in large layers of ice called **glaciers.** A smaller portion is found in groundwater, the water beneath the surface of the land. Only a tiny fraction of Earth's water is found in streams, lakes, wetlands, and the atmosphere.

Most of the fresh water on Earth is frozen in glaciers.

NEXT GENERATION SCIENCE STANDARDS | DISCIPLINARY CORE IDEAS
ESS2.C: The Roles of Water in Earth's Surface Processes
Nearly all of Earth's available water is in the ocean. Most fresh water is in glaciers or underground; only a tiny fraction is in streams, lakes, wetlands, and the atmosphere. (5-ESS2-2)

Today many people are working together to clean up the air. Most factories in the United States now have devices in their smokestacks that remove some of the pollution before it is released into the air. Planting trees and rooftop gardens in cities also helps reduce air pollution.

This rooftop garden not only helps clean the air, it also provides fresh food for a restaurant.

Wrap It Up!

1. **Recall** What are three ways that burning fuels affects the quality of the air?

2. **Explain** What are some ways that people are working together to reduce air pollution?

3. **Apply** Instead of riding in a car, you decide to ride your bicycle to school. How could this decision affect air quality? Explain.

Humans Impact Space

You might think that the space around Earth is empty. But instead, it contains millions of pieces of human-made trash—**space junk**! For more than 50 years, humans have been sending satellites into orbit around Earth. When satellites stop working, they become space junk. In addition to old satellites, space junk includes pieces of rockets, metal shards from objects that collided in space, and even tools left behind by astronauts.

Space junk is still in orbit but is no longer controlled by people on Earth. The objects are moving rapidly—27,400 kilometers per hour (17,000 miles per hour) or more! These objects can cause serious damage if they hit a satellite, spacecraft, or an astronaut.

Eventually Earth's gravity pulls bits of space junk back into Earth's atmosphere. Most of them burn up before reaching Earth's surface. Space junk that reaches the surface usually falls into the ocean. No one has ever been injured by space junk falling to Earth.

Today space scientists are working to reduce the amount of trash that is left in space. They are using radar to monitor the location of large objects. They use this information to help spacecraft, such as the International Space Station, avoid collisions with space junk.

NEXT GENERATION SCIENCE STANDARDS | DISCIPLINARY CORE IDEAS
ESS3.C: Human Impacts on Earth Systems
Human activities in agriculture, industry, and everyday life have had major effects on the land, vegetation, streams, ocean, air, and even outer space. But individuals and communities are doing things to help protect Earth's resources and environments. (5-ESS3-1)

1 Observe each object, including its shape, size, weight, and what it is made of. Predict how Earth's gravitational force will affect each object as you drop it on a table, as you gently push it off the edge of a desk or table, and as you gently toss it in the air. Record your predictions in your science notebook.

2 Stand by a desk or table. Take turns with a partner. Drop one object from about shoulder height onto the desk or table. Record your observations. Repeat this for every object.

3 Gently push each object off of the edge of the desk or table. Record your observations in your chart. Repeat this for each object. Then compare your predictions with your observations.

4 Gently toss one object slightly up in the air. Do not toss it in your partner's direction. Record your observations. Repeat this for every object.

Wrap It Up!

1. **Compare** Did your predictions support your results? Why do you think they were the same or different?

2. **Support an Argument** Use evidence from your investigation to support an argument that the force of Earth's gravity on an object is directed down.

Earth, Sun, and Moon

Two of Earth's close neighbors in space—the moon and the sun—form a system with Earth. They move independently and as a system. All three **revolve,** or move around other objects. Earth revolves around the sun, and—at the same time— the moon revolves around Earth. The sun, Earth, and moon together revolve around the center of our galaxy, the Milky Way.

The gravitational force between the sun and Earth keeps Earth in motion around the sun.

Sun

NEXT GENERATION SCIENCE STANDARDS | DISCIPLINARY CORE IDEAS
PS2.B: Types of Interactions
The gravitational force of Earth acting on an object near Earth's surface pulls that object toward the planet's center. (5-PS2-1)

The motion of one object around another is called **revolution.** The sun, Earth, and moon revolve as the result of **gravitational force.** This force pulls objects toward each other. The gravitational force between Earth and the less-massive moon keeps the moon moving around Earth. The gravitational force between Earth and the sun keeps Earth revolving around this massive star.

The gravitational force between Earth and the moon keeps the moon in motion around Earth.

Moon

Earth

is diagram is not drawn to scale. e sun's diameter is more than 9 times larger than the diameter Earth. The actual distances tween the moon, Earth, and the n is far greater than shown.

Wrap It Up!

1. **Define** What is revolution?

2. **Explain** Tell why and how Earth, the moon, and the sun revolve.

3. **Infer** Gravitational force is related to mass. Infer which has a greater gravitational pull on Earth—the sun or the moon. Explain.

Our Star—the Sun

A **star** is a ball of hot gases that gives off light and other types of energy. Stars range greatly in size and in the amounts of light and energy they give off. Even though the sun appears to be quite large to us on Earth, it is really a medium-sized star.

The sun appears larger and brighter than other stars in the sky because it is the closest star to Earth. It is so close, in fact, that its brightness blocks out the light of larger and brighter stars that are farther away. That's why you can't see other stars during daylight hours.

The sun is closer to Earth than the other stars. That makes its light look brighter than stars that are farther away.

NEXT GENERATION SCIENCE STANDARDS | DISCIPLINARY CORE IDEAS
ESS1.A: The Universe and Its Stars
The sun is a star that appears larger and brighter than other stars because it is closer.

Even stars that are bigger and brighter than the sun are visible only at night.

Size and Distance

1 Put an object on your desk or other surface. Measure the object's length, or its distance across the middle if it is round. Record your measurement.

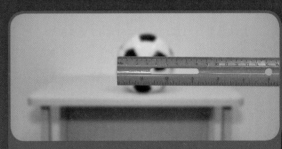

2 Move at least 2 m away from the object. Hold up your ruler to measure the apparent size of the object from your new position. Record your measurement.

? How did the size of the object appear to differ when you viewed it from different distances?

my science notebook

Wrap It Up!

1. **Describe** What is a star?

2. **Explain** Why does the sun seem to be brighter and larger than other stars in the sky?

161

Apparent Brightness

? **How does a light's brightness appear to change with distance?**

You know the sun appears larger and brighter than other stars in the sky. That is because it is the closest star to Earth. In this investigation, you'll observe how distance affects brightness. Can you make a light from a flashlight appear brighter? Can you make it appear less bright? You will use your observations to support an argument about the cause of differences in the apparent brightness of the sun compared to other stars.

Each star's size and distance from Earth affects how bright it appears to us.

Materials

3 penlights	tape	tissue paper

meterstick

NEXT GENERATION SCIENCE STANDARDS | PERFORMANCE EXPECTATION
5-ESS1-1. Support an argument that differences in the apparent brightness of the sun compared to other stars is due to their relative distances from the Earth.

1 Make a chart for your data. Use the tape and a pencil to label the penlights as **A**, **B**, and **C**.

Trial	Penlight	Distance from Observer	Observed Brightness
	A	2m	
1	B	2m	
	C	4m	
	A	4m	
	B	2m	
2	C	6m	

2 Use the meterstick to measure three distances—2 m, 4 m, and 6 m—from a wall, desk, or table. Mark and label each distance with a piece of tape.

3 **TRIAL 1:** Have each person cover a penlight with a piece of tissue paper and stand at the 4-m mark. Tell them to turn on the penlights. Observe and record brightness using these words: *very bright, bright,* or *dim.*

4 **TRIAL 2:** Repeat step 3 with one person at the 4-m mark, one at the 2-m mark, and one at the 6-m mark. The person at the 6-m mark should be closest to you. Rank and record the brightness of each light as *very bright, bright,* or *dim.*

Wrap It Up!

1. **Compare** Describe the brightness of the model stars in Trial 1.

2. **Infer** Which of the model stars in Trial 2 could represent the sun? Explain.

3. **Support an Argument** Why can stars with the same brightness appear dimmer or brighter than they actually are? Use your observations to support your argument.

Day and Night

Think about what happens when you **rotate,** or spin around. As you turn, you face different directions and see different things. Places on Earth are like that. Earth does not stand still. It rotates on an imaginary line called an **axis** that runs through the North and South Poles.

Half of Earth always faces the sun and is lit by sunlight. There it is day. The other half of Earth faces away from the sun and is dark. There it is night. Earth makes one complete rotation approximately every 24 hours. That's why one complete day–night cycle lasts 24 hours.

In about 24 hours, most areas on the surface of Earth rotate into sunlight, then into darkness, and then back into sunlight.

NEXT GENERATION SCIENCE STANDARDS | DISCIPLINARY CORE IDEAS
ESS1.B: Earth and the Solar System
The orbits of Earth around the sun and of the moon around Earth, together with the rotation of Earth about an axis between its North and South poles, cause observable patterns. These include day and night; daily changes in the length and direction of shadows; and different positions of the sun, moon, and stars at different times of the day, month, and year. (5-ESS1-2)

1 Put the poster board in a sunny place. Put a clay ball on the poster board. Push the end of a pencil into the clay.

2 Mark an *X* to show the direction of the sun. Record how high the sun looks in the sky. Trace the pencil's shadow. Write the date and time next to the shadow outline and the *X*.

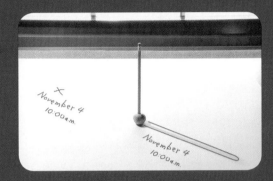

My science notebook

3 Use a meterstick to measure the shadow. Record the measurement and the time in your science notebook. Then repeat steps 2 and 3 at three more times during the day.

4 Use a colored pencil to draw where you predict the shadow will be in 1 hour. After 1 hour, repeat steps 2 and 3 one more time.

The location of the rider's shadow is determined by the position of the sun in the sky.

My science notebook

Wrap It Up!

1. **Identify** What patterns in length and movement did you observe with the shadows?

2. **Explain** Did your results support your prediction? How is the sun's position related to the position and length of the shadows?

Revolution and the Seasons

You can't feel it, but Earth moves. It rotates around its imaginary axis, which is tilted at an angle. Earth also revolves, or follows a regular path, around the sun. The path a revolving body follows is its **orbit.** Earth's tilt on its axis and its revolution around the sun cause observable patterns called seasons. Weather and the number of daylight hours at most places on Earth change with the seasons.

Earth's tilt on its axis also causes the seasons to be opposite in the Northern and Southern Hemispheres. For example, when it is summer in the Northern Hemisphere, it is winter in the Southern Hemisphere. Study the diagrams to see how Earth's changing position causes seasons. The captions describe seasonal changes in the Northern Hemisphere. Note that the sizes of Earth and the sun and the distance between them are not to scale in the drawings.

Earth's rotation and its revolution around the sun produce observable patterns.

These diagrams are not drawn to scale. The sun's diameter is many times larger than the diameter of Earth. The actual distance between Earth and the sun is far greater than shown.

NEXT GENERATION SCIENCE STANDARDS | DISCIPLINARY CORE IDEAS
ESS1.B: Earth and the Solar System
The orbits of Earth around the sun and of the moon around Earth, together with the rotation of Earth about an axis between its North and South poles, cause observable patterns. These include day and night; daily changes in the length and direction of shadows; and different positions of the sun, moon, and stars at different times of the day, month, and year. (5-ESS1-2)

As you look outward from the school, you can see only certain objects on certain sides of the school. Now think about Earth orbiting the sun. Earth's position changes throughout the year. Certain stars are visible only because we are looking outward into a different part of the night sky that is no longer blotted out by the sun's light.

MID-NOVEMBER
9:00 p.m., looking south

EARLY MARCH
9:00 p.m., looking south

Orion the hunter is visible in the night sky from late fall through early spring. In fall Orion is seen in the eastern sky. By spring it appears in the western sky.

Summer You can see Lyra the harp in the dark summer sky.

This diagram is not drawn to scale. The sun's diameter is many times larger than the diameter of Earth. The actual distance between Earth and the sun is far greater than shown.

Wrap It Up!

1. **Explain** Why are most stars visible only at night?

2. **Cause and Effect** Why do some constellations seem to change during the year?

Represent Data

Remember the constellation Orion? It moves through the sky from late fall through early spring. In fall, if you look south at 9:00 p.m., you'll see Orion in the eastern sky. In December it is high overhead. By spring it is visible in the western sky.

You can apply what you know about Orion to other constellations. Study the diagram. Note the constellation listed for each season. These constellations would be high overhead in the night sky in the identified seasons. After you study the diagram, answer the questions.

1. **Identify.**
   Which constellation is high in the sky in spring? In summer? In fall? In winter?

2. **Represent data in graphical displays.** *My science notebook*
   Choose one of the constellations shown. Draw a series of three illustrations. First, show where an observer would see the constellation early in the season. Second, show where an observer would see the constellation mid-season. Third, show where an observer would see the constellation late in the season. Then use your graphical displays to describe the pattern of the appearance of stars in the sky.

NEXT GENERATION SCIENCE STANDARDS | PERFORMANCE EXPECTATION
5-ESS1-2. Represent data in graphical displays to reveal patterns of daily changes in length and direction of shadows, day and night, and the seasonal appearance of some stars in the night sky.

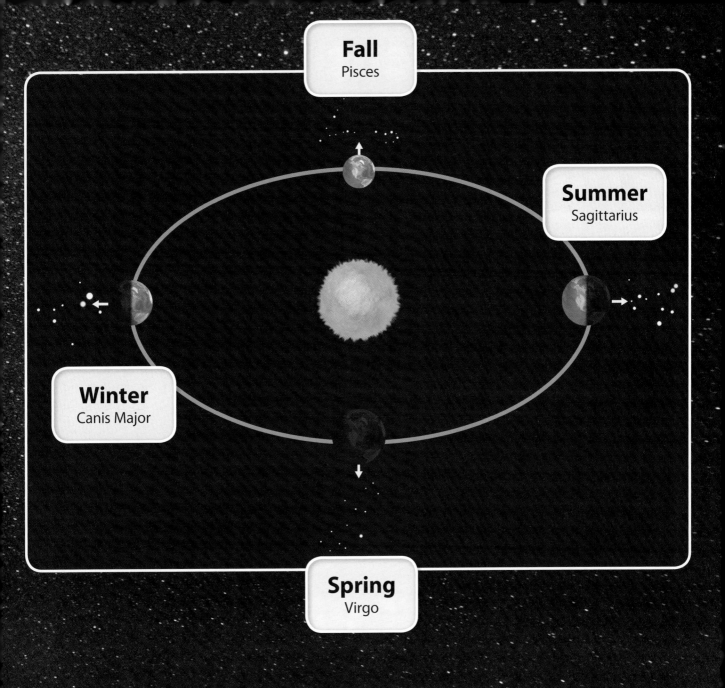

Fall
Pisces

Summer
Sagittarius

Winter
Canis Major

Spring
Virgo

Which constellations are visible and where they appear in the sky depends on the time of year.

Moon Motions

Like Earth, the moon rotates, or spins, around an imaginary axis. The moon's orbit, however, is around Earth. As Earth completes its revolution around the sun once each year, the moon revolves around Earth about twelve times.

Motions of Earth and the moon result in observational patterns. Earth's rotation causes the moon to appear to

The apparent shape of the moon changes throughout the month in a repeating pattern.

NEXT GENERATION SCIENCE STANDARDS | DISCIPLINARY CORE IDEAS
ESS1.B: Earth and the Solar System
The orbits of Earth around the sun and of the moon around Earth, together with the rotation of Earth about an axis between its North and South poles, cause observable patterns. These include day and night; daily changes in the length and direction of shadows; and different positions of the sun, moon, and stars at different times of the day, month, and year. (5-ESS1-2)

rise toward the east, move across the sky, and set toward the west. The revolution of the moon around Earth causes the moon's shape to seem to change from day to day. The revolution of the moon around Earth is a key reason that the moon's apparent change in shape are patterns that occur again and again.

The moon isn't only visible at night. Much of the month, you can see the moon in the daytime, too.

Wrap It Up!

my science notebook

1. **Contrast** Describe the difference between the terms *rotate* and *revolve*.

2. **Describe** Tell about the two types of motion of Earth's moon.

3. **Summarize** Explain why the moon seems to move across the sky and why its shape seems to change during the month.

As you study the pictures, you'll see that some days the moon looks like a full circle. Other days it looks like part of a circle. But the moon doesn't change shape. It is always shaped like a ball. What changes is how much of

A

NEW MOON

B

WAXING CRESCENT MOON

C

FIRST QUARTER MOON

D

WAXING GIBBOUS MOON

the moon's lighted half you can see from Earth. Compare the phase pictures with the diagram on the next page. Like Earth, half of the moon always faces the sun. That side is lit. How much of this lit side is visible depends on where the moon is in its orbit around Earth.

NEXT GENERATION SCIENCE STANDARDS | DISCIPLINARY CORE IDEAS
ESS1.B: Earth and the Solar System
The orbits of Earth around the sun and of the moon around Earth, together with the rotation of Earth about an axis between its North and South poles, cause observable patterns. These include day and night; daily changes in the length and direction of shadows; and different positions of the sun, moon, and stars at different times of the day, month, and year. (5-ESS1-2)

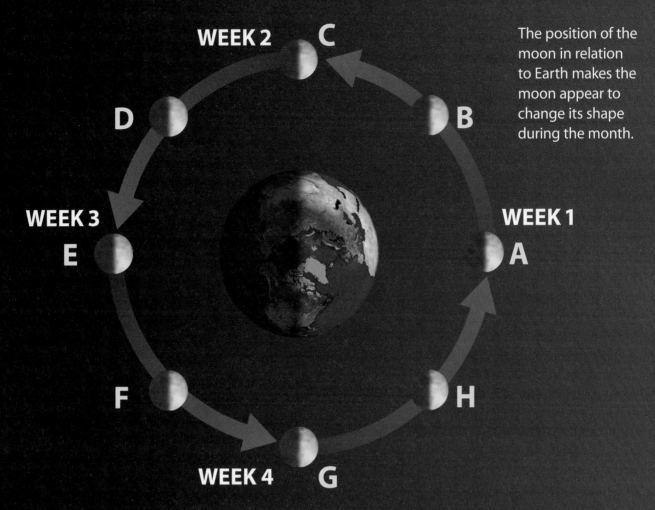

WEEK 2 C

D

The position of the moon in relation to Earth makes the moon appear to change its shape during the month.

B

WEEK 3
E

WEEK 1
A

F

H

WEEK 4 G

E
FULL MOON

F
WANING GIBBOUS MOON

G
THIRD QUARTER MOON

H
WANING CRESCENT MOON

Earth's moon appears to change shape from day to day. These changes in shape are called moon phases.

Wrap It Up!

my science notebook

1. **Sequence** Name the eight phases of the moon, beginning with the new moon.

2. **Describe** Tell how the moon seems to change shape as it goes from new moon to full moon, and from full moon back to new moon.

3. **Summarize** Explain why the moon seems to change its shape during the month.

Moon Phases

? **How does the moon's orbit affect how it looks at different times of the month?**

The moon's orbit around Earth results in different phases. In this investigation, you'll model how the moon goes through its phases.

The moon phases calendar shows how the appearance of the moon changes a little bit each day.

MOON PHASES CALENDAR

Sunday	Monday	Tuesday	Wednesday	Thursday	Friday	Saturday
1	2	3	4	5	6	7
8	9	10	11	12	13	14
15	16	17	18	19	20	21
22	23	24	25	26	27	28

Materials

lamp	ball	craft stick

meterstick

NEXT GENERATION SCIENCE STANDARDS | DISCIPLINARY CORE IDEAS
ESS1.B: Earth and the Solar System
The orbits of Earth around the sun and of the moon around Earth, together with the rotation of Earth about an axis between its North and South poles, cause observable patterns. These include day and night; daily changes in the length and direction of shadows; and different positions of the sun, moon, and stars at different times of the day, month, and year. (5-ESS1-2)

conservation of matter
(kon-sur-VĀ-shun uv MA-tur)
Conservation of matter is a principle that states that the amount of matter does not increase or decrease after a reaction or change of state. (p. 29)

constellation (kon-sti-LĀ-shun)
A constellation is a group of stars that forms a particular shape in the sky and has been given a name. (p. 174)

consumer (kun-SŪ-mur)
A consumer is a living thing that eats plants or animals. (p. 67)

coral reef (KOR-ul RĒF)
A coral reef is an ocean ridge made up of coral skeletons and living coral. (p. 113)

D

decomposer (dē-kum-PŌZ-ur)
A decomposer is an organism that breaks down dead organisms and the waste of living things. (p. 74)

deforestation (dē-for-is-TĀ-shun)
Deforestation is the cutting down or burning of all the trees in an area. (p. 132)

delta (DEL-tuh)
A delta is new land that forms at the mouth of a river. (p. 92)

deposition (de-pō-ZI-shun)
Deposition is the laying down of sediment and rock in a new place. (p. 115)

dissolve (di-ZAHLV)
To dissolve means to move particles of a solid throughout a liquid to form a solution. (p. 8)

E

ecosystem (Ē-kō-sis-tum)
An ecosystem is all the living things and nonliving things in an area and their interactions. (p. 79)

electrical conductor (i-LEK-tri-kul kon-DUK-ter)
An electrical conductor is a material through which electric energy can flow easily. (p. 20)

electrical conductivity
(i-LEK-tri-kul kon-duk-TIV-i-tē)
Electrical conductivity is a measure of how well electricity can move through a material. (p. 20)

electrical insulator (i-LEK-trik-ul IN-su-lā-tur)
An electrical insulator is a material that slows or stops the flow of electricity. (p. 21)

erosion (i-RŌ-zhun)
Erosion is the movement of rocks or soil caused by wind, water, or ice. (p. 115)

evaporate (i-VAP-uh-rāt)
To evaporate means to change from a liquid to a gas. (pp. 8, 103)

F

food web (FŪD web)
A food web is a network of food chains that shows how energy moves through an ecosystem. (p. 72)

fungi (FUN-gī)
Fungi are types of plants, such as molds and mushrooms, that have no chlorophyll and live on dead and decaying things. Fungi are decomposers. (p. 74)

G

Gas (GAS)
gas is matter that spreads to fill a space. (p. 6)

geosphere (JĒ-ō-sfear)
The geosphere is the solid outer part of Earth composed of rock and thought to be about 60 miles thick. (p. 98)

glacier (GLĀ-shur)
A glacier is a huge, slow-moving mass of ice. (p. 124)

gravity (GRA-vi-tē)
Gravity is a force that pulls objects toward each other. (p. 154) *also* **gravitational force** (GRA-vi-TĀ-shun-ul FORS) (p. 159)

groundwater (GROWND-wah-tur)
Groundwater is water held underground in spaces within soil and rock. (p. 102)

H

hardness (HARD-nis)
Hardness is a measure of how resistant a material is to scratching, bending, or denting. (p. 14)

hydroponics (hī-drō-PAH-niks)
Hydroponics is a method of growing plants in water instead of soil. (p. 58)

hydrosphere (HĪ-drō-sfear)
The hydrosphere is all the water at or near Earth's surface, including liquid bodies of water, frozen water as ice and snow, water found underground, and water vapor in the atmosphere. (p. 98)

I

invasive species (in-VĀ-siv SPĒ-shēz)
An invasive species is a species that has been brought to a new place by people and can harm the environment. (p. 87)

L

liquid (LIK-wid)
A liquid is matter that takes the shape of its container. (p. 6)

M

magnetism (MAG-nuh-ti-zum)
Magnetism is a force between magnets and objects magnets attract. (p. 18)

Content Consultants

Randy L. Bell, Ph.D.
Associate Dean and Professor of
Science Education, College of
Education, Oregon State University

Malcolm B. Butler, Ph.D.
Associate Professor of Science
Education, School of Teaching,
Learning and Leadership, University of
Central Florida

Kathy Cabe Trundle, Ph.D.
Department Head and Professor,
STEM Education, North Carolina
State University

Judith S. Lederman, Ph.D.
Associate Professor and Director of
Teacher Education, Illinois Institute
of Technology

Acknowledgments
Grateful acknowledgment is given to the
authors, artists, photographers, museums,
publishers, and agents for permission to
reprint copyrighted material. Every effort
has been made to secure the appropriate
permission. If any omissions have been made
or if corrections are required, please contact
the Publisher.

NEXT GENERATION **SCIENCE** STANDARDS *For States, By States* is a registered trademark
of Achieve. Neither Achieve
nor the lead states and partners that
developed the Next Generation Science
Standards was involved in the production
of, and does not endorse, this product.

Photographic and Illustrator Credits
Front cover wrap ©William Saar/Alamy.
Back cover (t) ©Rebecca Hale/National
Geographic Creative. (b) ©Peter McBride/
National Geographic Creative.

Acknowledgments and credits continued on
page 206.

For permission to use material from this text
or product, submit all requests online at
www.cengage.com/permissions

Further permissions questions can be
emailed to permissionrequest@
cengage.com

Visit National Geographic Learning online at
NGL.Cengage.com

Visit our corporate website at
www.cengage.com

Printed in the USA.
RR Donnelley

ISBN: 978-12858-46378

16 17 18 19 20 21 22 23

10 9 8 7 6